WORKSHEETS
WITH THE MATH COACH

GEX, INCORPORATED

INTERMEDIATE ALGEBRA
SEVENTH EDITION

John Tobey
North Shore Community College

Jeffrey Slater
North Shore Community College

Jamie Blair
Orange Coast College

Jennifer Crawford
Normandale Community College

PEARSON

Boston Columbus Indianapolis New York San Francisco Upper Saddle River
Amsterdam Cape Town Dubai London Madrid Milan Munich Paris Montreal Toronto
Delhi Mexico City Sao Paulo Sydney Hong Kong Seoul Singapore Taipei Tokyo

The author and publisher of this book have used their best efforts in preparing this book. These efforts include the development, research, and testing of the theories and programs to determine their effectiveness. The author and publisher make no warranty of any kind, expressed or implied, with regard to these programs or the documentation contained in this book. The author and publisher shall not be liable in any event for incidental or consequential damages in connection with, or arising out of, the furnishing, performance, or use of these programs.

Reproduced by Pearson from electronic files supplied by the author.

Copyright © 2013, 2009, 2005 Pearson Education, Inc.
Publishing as Pearson, 75 Arlington Street, Boston, MA 02116.

All rights reserved. No part of this publication may be reproduced, stored in a retrieval system, or transmitted, in any form or by any means, electronic, mechanical, photocopying, recording, or otherwise, without the prior written permission of the publisher. Printed in the United States of America.

ISBN-13: 978-0-321-75901-6
ISBN-10: 0-321-75901-X

6 V011 16 15 14

www.pearsonhighered.com

PEARSON

Worksheets with the Math Coach

Intermediate Algebra, Seventh Edition

Table of Contents

Chapter 1 Basic Concepts
1.1 The Real Number System

Vocabulary
set • elements • infinite • empty set • natural numbers • whole numbers
integers • rational numbers • variables • terminating decimal • repeating decimal
irrational numbers • real numbers • subset • opposite • reciprocal • percent
commutative property • associative property • identity property • inverse property

1. The commutative, associative, identity, and inverse properties are some of the properties of _____.

2. A set that contains no elements is called the _____.

3. The set R of real numbers is the set of all rational or _____.

4. The _____ are the whole numbers plus the negatives of all natural numbers.

Example	Student Practice
1. Name the sets to which each of the following numbers belongs. **(a)** 5 This number is a natural number. Thus, we can say it is also a whole number, an integer, a rational number, and a real number. **(b)** 0.2666... This number is a repeating decimal. Thus, it is a rational number and a real number. **(c)** 1.4371826138526... It doesn't have any repeating pattern. It is an irrational number and a real number. **(d)** 0 This number is a whole number, an integer, a rational number, and a real number.	**2.** Name the sets to which each of the following numbers belongs. **(a)** 1.34 **(b)** $\frac{2}{11}$ **(c)** −6 **(d)** 7.627854...

Vocabulary Answers: 1. real numbers 2. empty set 3. irrational numbers 4. integers

Copyright © 2013 Pearson Education, Inc.

Example	Student Practice
3. State the name of the property that justifies each statement. **(a)** $5+(7+1)=(5+7)+1$ The left side of the equation has the last two numbers grouped. The right side has the first two numbers grouped. The associative property of addition justifies the statement. **(b)** $x+0.6=0.6+x$ The positions of the letter and number are swapped. The commutative property of addition justifies the statement.	**4.** State the name of the property that justifies each statement. **(a)** $z+6=6+z$ **(b)** $22+0=22$
5. State the name of the property that justifies each statement. **(a)** $5\cdot y=y\cdot 5$ The positions of the letter and number are swapped. The commutative property of multiplication justifies the statement. **(b)** $6\cdot\dfrac{1}{6}=1$ This product results in the identity element of multiplication. The inverse property of multiplication justifies the statement. **(c)** $5(9+3)=5\cdot 9+5\cdot 3$ The statement links multiplication with addition. The distributive property of multiplication over addition justifies the statement.	**6.** State the name of the property that justifies each statement. **(a)** $3\cdot(2\cdot z)=(3\cdot 2)z$ **(b)** $17(x+y)=17\cdot x+17\cdot y$ **(c)** $8\cdot\dfrac{1}{8}=1$

2

Copyright © 2013 Pearson Education, Inc.

Extra Practice

1. List the rational numbers in the following set.

$$\left\{-33, \ -\frac{48}{5}, \ -2.347, \ -\frac{\pi}{2}, \ -0.222..., \right.$$

$$\left. 0, \ \frac{1}{8}, \ \pi, \ \sqrt{81}, \ \sqrt{15}, \ 43.5 \right\}$$

2. List the whole numbers in the following set.

$$\left\{-33, \ -\frac{48}{5}, \ -2.347, \ -\frac{\pi}{2}, \ -0.222..., \right.$$

$$\left. 0, \ \frac{1}{8}, \ \pi, \ \sqrt{81}, \ \sqrt{15}, \ 43.5 \right\}$$

3. Name the property that justifies the statement. Any variable represents a real number.

$$5 \cdot (2 \cdot x) = (5 \cdot 2) \cdot x$$

4. Name the property that justifies the statement. Any variable represents a real number.

$$0 + y = y$$

Concept Check

Explain what properties would be needed to justify this statement.

$$(4)(3.5 + 9.3) = (9.3 + 3.5)(4)$$

Copyright © 2013 Pearson Education, Inc.

Copyright © 2013 Pearson Education, Inc.

Chapter 1 Basic Concepts
1.2 Operations with Real Numbers

Vocabulary
real number line   •   is less than   •   is greater than   •   absolute value
opposite   •   additive inverse   •   multiplication property of zero

1. The opposite of a number is also called the _____.

2. The _____ has positive numbers that lie to the right of zero and negative numbers that lie to the left of zero.

3. The _____ states that if b is any real number, then $b \cdot 0 = 0$ and $0 \cdot b = 0$.

4. The _____ of a number, x, is its distance from 0 on a number line.

Example	**Student Practice**
1. Evaluate.	**2.** Evaluate.
(a) $\lvert -8 \rvert$	**(a)** $\left\lvert -\dfrac{1}{9} \right\rvert$
The definition of the absolute value of x is as follows.	
$\lvert x \rvert = \begin{cases} x, & \text{if } x \geq 0 \\ -x, & \text{if } x < 0 \end{cases}$	
Notice that $-8 < 0$. Find $\lvert -8 \rvert$.	**(b)** $\lvert -5.7 \rvert$
$\lvert -8 \rvert = 8$	
(b) $\lvert 0 \rvert$	**(c)** $\lvert 17 - 4 \rvert$
$\lvert 0 \rvert = 0$	
(c) $\lvert 5 - 3 \rvert$	
$\lvert 5 - 3 \rvert = \lvert 2 \rvert = 2$	

Vocabulary Answers: 1. additive inverse 2. real number line 3. multiplication property of zero
4. absolute value

Copyright © 2013 Pearson Education, Inc.

Example	**Student Practice**
3. Add.	**4.** Add.
(a) $-5+(-0.6)$	**(a)** $-3.6+(-0.9)$
Rule 1.1 states that to add two real numbers with the same sign, add their absolute values. The sum takes the common sign.	
$-5+(-0.6)=-5.6$	
(b) $\dfrac{2}{5}+\dfrac{3}{7}$	**(b)** $\dfrac{1}{5}+\dfrac{1}{3}$
Apply Rule 1.1. Obtain a common denominator before adding.	
$\dfrac{2}{5}+\dfrac{3}{7}=\dfrac{14}{35}+\dfrac{15}{35}=\dfrac{29}{35}$	
5. Add.	**6.** Add.
(a) $12+(-5)$	**(a)** $(-19)+6$
Rule 1.2 states that to add two numbers with different signs, find the difference between their absolute values. The answer takes the sign of the number with the larger absolute value.	
$12+(-5)=7$	
(b) $-\dfrac{1}{3}+\dfrac{1}{4}$	**(b)** $-\dfrac{1}{2}+\dfrac{3}{4}$
Apply Rule 1.2.	
$-\dfrac{1}{3}+\dfrac{1}{4}=-\dfrac{4}{12}+\dfrac{3}{12}=-\dfrac{1}{12}$	

Copyright © 2013 Pearson Education, Inc.

Example	Student Practice
7. Use Rule 1.3 to subtract.	**8.** Use Rule 1.3 to subtract.
(a) $5-7$	**(a)** $-16-(-9)$
Rule 1.3 states that to subtract b from a, add the opposite (additive inverse) of b to a. Thus, $a-b=a+(-b)$.	
$5-7=5+(-7)=-2$	**(b)** $-\dfrac{3}{5}-\dfrac{1}{4}$
(b) $-12-(-3)$	
$-12-(-3)=-12+3=-9$	
(c) $-0.06-0.55$	**(c)** $2.7-(-1.9)$
$\begin{aligned} -0.06-0.55 &= -0.06+(-0.55) \\ &= -0.61 \end{aligned}$	
9. Evaluate. $\dfrac{1.6}{-0.08}$	**10.** Evaluate. $\left(-\dfrac{3}{16}\right)(5)$
Rule 1.4 states that when you multiply or divide two real numbers with different signs, the answer is a negative number.	
$\dfrac{1.6}{-0.08}=-20$	
11. Evaluate. $\dfrac{-3}{5}\cdot\dfrac{-2}{11}$	**12.** Evaluate. $\dfrac{4}{9}\div 3$
Rule 1.5 states that when you multiply or divide two real numbers with like signs, the answer is a positive number.	
$\dfrac{-3}{5}\cdot\dfrac{-2}{11}=\dfrac{(-3)(-2)}{(5)(11)}=\dfrac{6}{11}$	

Copyright © 2013 Pearson Education, Inc.

Example	Student Practice
13. Evaluate. $20 \div (-4) \times 3 + 2 + 6 \times 5$	**14.** Evaluate. $15 \div 3 - 7(-2) + 3 + 5(4)$

13. (continued)

Beginning at the left, we do the multiplication and division as we encounter it. Here we encounter division first and then multiplication.

$$20 \div (-4) \times 3 + 2 + 6 \times 5 = -5 \times 3 + 2 + 6 \times 5$$
$$= -15 + 2 + 6 \times 5$$
$$= -15 + 2 + 30$$

Then add and subtract from left to right.

$$-15 + 2 + 30 = -13 + 30 = 17$$

15. Evaluate. $\dfrac{13 - (-3)}{5(-2) - 6(-3)}$	**16.** Evaluate. $\dfrac{-22 - (6)}{-7(-3) + 6(-3) + 4}$

Complete the operations in the numerator and complete the operations in the denominator before simplifying.

$$\frac{13 - (-3)}{5(-2) - 6(-3)} = \frac{13 + 3}{-10 + 18} = \frac{16}{8} = 2$$

Extra Practice

1. Evaluate. $|3 - 7|$

2. Evaluate. $\left(-\dfrac{4}{7}\right) + \left(\dfrac{3}{5}\right)$

3. Evaluate. $\left(\dfrac{7}{8}\right) \div \left(-\dfrac{21}{32}\right)$

4. Evaluate. $\dfrac{6 + (-6)}{22}$

Concept Check

Explain what steps need to be taken in what order to perform the following operations.

$$\frac{3(-2) + 8}{5 - 9}$$

Copyright © 2013 Pearson Education, Inc.

Name: _____   Date: _____

Instructor: _____   Section: _____

Chapter 1 Basic Concepts
1.3 Powers, Square Roots, and the Order of Operations

Vocabulary
exponent • power • base • exponential notation • principal square root
radical • radicand • square root • perfect square • order of operations

1. The number or expression under the radical sign is called the _____.

2. The _____ is the nonnegative square root of a number.

3. Exponents or _____ are used to indicate repeated multiplication of a base.

4. When many arithmetic operations or grouping symbols are used, use the _____.

Example	Student Practice
1. Write in exponential notation. $(-2)(-2)(-2)(-2)(-2)$ Use the fact that $x^n = \underbrace{x \cdot x \cdot x \cdot x \cdots}_{n \text{ factors}}$ when x is a real number and n is a positive integer. $(-2)(-2)(-2)(-2)(-2) = (-2)^5$	**2.** Write in exponential notation. $a \cdot a \cdot a \cdot a \cdot a \cdot a$
3. Evaluate. **(a)** $(-2)^4$ Note that we are raising -2 to the fourth power. The parentheses are used to show the base is negative. $(-2)^4 = (-2)(-2)(-2)(-2) = 16$ **(b)** -2^4 Here the base is 2, not -2. Find the negative of 2^4. $-2^4 = -(2 \cdot 2 \cdot 2 \cdot 2) = -16$	**4.** Evaluate. **(a)** $\left(-\dfrac{1}{4}\right)^3$ **(b)** -4^6

Vocabulary Answers: 1. radicand 2. principal square root 3. powers 4. order of operations

9

Copyright © 2013 Pearson Education, Inc.

Example	Student Practice
5. Find the square roots of 25. What is the principal square root? Since $(-5)^2 = 25$ and $5^2 = 25$, the square roots of 25 are 5 and -5. The principal square root is 5.	**6.** Find the square roots of 64. What is the principal square root?
7. Evaluate. (a) $\sqrt{81}$ $\sqrt{81} = 9$ because $9^2 = 81$. (b) $\sqrt{0}$ $\sqrt{0} = 0$ because $0^2 = 0$. (c) $-\sqrt{49}$ $-\sqrt{49} = -\left(\sqrt{49}\right) = -(7) = -7$	**8.** Evaluate. (a) $\sqrt{144}$ (b) $\sqrt{4}$ (c) $-\sqrt{36}$
9. Evaluate. (a) $\sqrt{0.04}$ $(0.2)^2 = (0.2)(0.2) = 0.04.$ Therefore, $\sqrt{0.04} = 0.2$. (b) $\sqrt{\dfrac{25}{36}}$ $\sqrt{\dfrac{25}{36}} = \dfrac{\sqrt{25}}{\sqrt{36}} = \dfrac{5}{6}$ Thus, $\sqrt{\dfrac{25}{36}} = \dfrac{5}{6}$. (c) $\sqrt{-16}$ This is not a real number.	**10.** Evaluate. (a) $\sqrt{0.49}$ (b) $\sqrt{\dfrac{81}{100}}$ (c) $\sqrt{-121}$

Copyright © 2013 Pearson Education, Inc.

Example	Student Practice
11. Evaluate. $2(8+7)-12$	**12.** Evaluate. $4-3\left[(3-4)+14\right]$

Work inside the parentheses first, then multiply. Finally, subtract.

$$2(8+7)-12 = 2(15)-12$$
$$= 30-12$$
$$= 18$$

13. Evaluate. $(4-6)^3+5(-4)+3$	**14.** Evaluate. $(3+2)^3-7(-4)+2$

Combine $4-6$ in parentheses then evaluate the exponent.

$$(4-6)^3+5(-4)+3 = (-2)^3+5(-4)+3$$
$$= -8+5(-4)+3$$

Multiply $5(-4)$.

$$-8+5(-4)+3 = -8-20+3$$

Combine $-8-20+3$.

$$-8-20+3 = -25$$

15. Evaluate. $2+66\div11\cdot3+2\sqrt{36}$	**16.** Evaluate. $3\sqrt{9}-2+48\div6+5$

Evaluate the square root first.

$$2+66\div11\cdot3+2\sqrt{36}$$
$$= 2+66\div11\cdot3+2\cdot6$$

Divide, then multiply.

$$2+66\div11\cdot3+2\cdot6 = 2+6\cdot3+2\cdot6$$
$$= 2+18+12$$

Finally, add.

$$2+18+12 = 32$$

Copyright © 2013 Pearson Education, Inc.

Extra Practice

1. Evaluate. $\left(-\dfrac{1}{2}\right)^5$

2. Find the principle square root. $-\sqrt{0.16}$

3. Evaluate. $2\left[7^2 + 4(10+4)\right]$

4. Evaluate. $\dfrac{8(5-2)+8\cdot 5}{8(6-4)}$

Concept Check

Explain what operations need to be done in what order to evaluate the following.

$$\dfrac{\sqrt{(-3)^3 - 6(-2)+15}}{|3-5|}$$

Copyright © 2013 Pearson Education, Inc.

Chapter 1 Basic Concepts
1.4 Integer Exponents and Scientific Notation

Vocabulary
negative exponent • product rule of exponents • quotient rule of exponents
zero power • power rules • rule of negative exponents • scientific notation

1. A positive number written in _____ has the form $a \times 10^n$, where $1 \le a < 10$ and n is an integer.

2. A positive exponent indicates repeated multiplication, a(n) _____ indicates repeated division.

3. Any nonzero real number raised to the _____ is equal to 1.

4. The _____ states that if x is a real number and n and m are integers, then $x^m \cdot x^n = x^{m+n}$.

Example	**Student Practice**
1. Simplify 2^{-5}. Do not leave negative exponents in your answers. If x is any nonzero real number and n is an integer, $x^{-n} = \dfrac{1}{x^n}$. $2^{-5} = \dfrac{1}{2^5} = \dfrac{1}{32}$	**2.** Simplify b^{-7}. Do not leave negative exponents in your answers.
3. Multiply. $\left(5x^2 y\right)\left(-2xy^3\right)$ Use the product rule of exponents, $x^m \cdot x^n = x^{m+n}$, where x is a real number and n and m are integers. $\left(5x^2 y\right)\left(-2xy^3\right)$ $= (5)(-2)\left(x^2 \cdot x^1\right)\left(y^1 \cdot y^3\right)$ $= -10x^3 y^4$	**4.** Multiply. $\left(7xy\right)\left(-3x^4 y^3\right)$

Vocabulary Answers: 1. scientific notation 2. negative exponent 3. zero power 4. product rule of exponents

Copyright © 2013 Pearson Education, Inc.

Example	Student Practice
5. Simplify. Do not leave negative exponents in your answers.	**6.** Simplify. Do not leave negative exponents in your answers.
(a) $3x^0$	**(a)** $-6x^0$
Use the rule that for any nonzero real number x, $x^0 = 1$.	
$3x^0 = 3(1) = 3$	
(b) $(3x)^0$	**(b)** $(4ab)^0$
Notice that the entire expression is raised to the zero power.	
$(3x)^0 = 1$	
(c) $(-2x^{-5})(y^3)^0$	**(c)** $(-3z^{-4})(y^2)^0$
$(-2x^{-5})(y^3)^0 = (-2x^{-5})(1)$	
$= (-2)\left(\dfrac{1}{x^5}\right) = \dfrac{-2}{x^5}$	
7. Divide. Then simplify your answer. $\dfrac{3x^{-5}y^{-6}}{27x^2y^{-8}}$	**8.** Divide. Then simplify your answer. $\dfrac{4x^{-3}y^{-5}}{24x^{-2}y^8}$
Use the quotient rule for exponents. This rule states that if x is a nonzero real number and n and m are integers, $\dfrac{x^m}{x^n} = x^{m-n}$.	
$\dfrac{3x^{-5}y^{-6}}{27x^2y^{-8}} = \dfrac{1}{9}x^{-5-2}y^{-6-(-8)}$	
$= \dfrac{1}{9}x^{-5-2}y^{-6+8} = \dfrac{1}{9}x^{-7}y^2 = \dfrac{y^2}{9x^7}$	

Copyright © 2013 Pearson Education, Inc.

Example	Student Practice
9. Simplify. $\left(\dfrac{5xy^{-3}}{2x^{-4}yz^{-3}}\right)^{-2}$	**10.** Simplify. $\left(\dfrac{2x^{-2}y^{-4}}{3x^{-4}y^{-6}z^{-3}}\right)^{-2}$

First remove the parentheses by using the power rules of exponents.

$$\left(\frac{5xy^{-3}}{2x^{-4}yz^{-3}}\right)^{-2} = \frac{5^{-2}x^{-2}y^{6}}{2^{-2}x^{8}y^{-2}z^{6}}$$

$$= \frac{2^{2}y^{6}y^{2}}{5^{2}x^{8}x^{2}z^{6}} = \frac{4y^{8}}{25x^{10}z^{6}}$$

---

11. Simplify $\left(-3x^{2}\right)^{-2}\left(2x^{3}y^{-2}\right)^{3}$. Express your answer with positive exponents only.

$$\left(-3x^{2}\right)^{-2}\left(2x^{3}y^{-2}\right)^{3} = (-3)^{-2}x^{-4} \cdot 2^{3}x^{9}y^{-6}$$

$$= \frac{2^{3}x^{9}}{(-3)^{2}x^{4}y^{6}}$$

$$= \frac{8x^{9}}{9x^{4}y^{6}} = \frac{8x^{5}}{9y^{6}}$$

12. Simplify $\left(5x^{2}\right)^{2}\left(-6x^{3}y^{-4}\right)^{-3}$. Express your answer with positive exponents only.

---

13. Write in decimal from.

(a) 8.8632×10^{4}

To change from scientific notation to decimal form, move the decimal point to the right or to the left the number of places indicated by the power of 10. For this problem, move the decimal point 4 places to the right.

$$8.8632\times10^{4} = 88,632$$

(b) 6.032×10^{-2}

Move the decimal point 2 places to the left, $6.032\times10^{-2} = 0.06032$.

14. Write in decimal from.

(a) 9.763×10^{2}

(b) 1.112×10^{-6}

Copyright © 2013 Pearson Education, Inc.

Example	Student Practice
15. Evaluate using scientific notation. $$\dfrac{(0.000000036)(0.002)}{0.000012}$$ Rewrite the expression using scientific notation. $$\dfrac{(3.6\times10^{-8})(2\times10^{-3})}{1.2\times10^{-5}}$$ Rewrite using the commutative property. Then simplify and use the rules of exponents. $$\dfrac{\overset{2}{\cancel{(3.6)}}(2)(10^{-8})(10^{-3})}{\underset{1}{\cancel{(1.2)}}(10^{-5})}=\dfrac{6.0}{1}\times\dfrac{10^{-11}}{10^{-5}}$$ $$=6.0\times10^{-11-(-5)}$$ $$=6.0\times10^{-6}$$	**16.** Evaluate using scientific notation. $$\dfrac{(0.000042)(0.002)}{0.0021}$$

Extra Practice

1. Simplify. Express your answer with positive exponents only.
$$(-3a^2b^3)(2a^4b^2)(-ab)$$

2. Simplify. Express your answer with positive exponents only.
$$\dfrac{4a^{24}}{8a^{36}}$$

3. Simplify. Express your answer with positive exponents only.
$$\left(\dfrac{4x^2y^3}{x^{-3}y^4}\right)^3$$

4. Write in scientific notation.
0.000032

Concept Check

Explain how you would simplify the following. $(3x^2y^{-3})(2x^4y^2)$

Copyright © 2013 Pearson Education, Inc.

Chapter 1 Basic Concepts
1.5 Operations with Variables and Grouping Symbols

Vocabulary
algebraic expression • term • coefficient • like terms • polynomial

1. A variable expression that contains terms with nonnegative integer exponents is called a(n) _____.

2. A(n) _____ is a real number, a variable, or a product or quotient of numbers and variables.

3. Any factor in a term is the _____ of the product of the remaining factors.

4. A collection of numerical values, variables and operation signs is called a(n) _____.

Example	Student Practice
1. List the terms in each algebraic expression.	**2.** List the terms in each algebraic expression.
(a) $5x + 3y^2$	**(a)** $8z^4 - 3y$
$5x$ is a product of a real number (5) and a variable (x), so $5x$ is a term. Similarly, $3y^2$ is a term.	
(b) $5x^2 - 3xy - 7$	**(b)** $3z + 4w + 7$
Rewrite $5x^2 - 3xy - 7$ using plus signs to help identify the terms.	
$5x^2 + (-3xy) + (-7)$	
The terms are $5x^2$, $(-3xy)$, and (-7).	

Vocabulary Answers: 1. polynomial 2. term 3. coefficient 4. algebraic expression

Copyright © 2013 Pearson Education, Inc.

Example	Student Practice
3. Identify the coefficient of each term. $5x^2 - 2x + 3xy$ The coefficient of the x^2 term is 5. The coefficient of the x term is -2. The coefficient of the xy term is 3.	**4.** Identify the coefficient of each term. $\dfrac{5}{6}x^2 - 2.5z + 18,000p^3$
5. Combine like terms. **(a)** $7x^2 - 2x - 8 + x^2 + 5x - 12$ Remember that the coefficient of x^2 is 1, so you are adding $7x^2 + 1x^2$. $7x^2 - 2x - 8 + x^2 + 5x - 12$ $= (7+1)x^2 + (-2+5)x + (-8-12)$ $= 8x^2 + 3x - 20$ **(b)** $\dfrac{1}{3}x^2 + \dfrac{1}{4}x - \dfrac{1}{6}x^2$ $\dfrac{1}{3}x^2 + \dfrac{1}{4}x - \dfrac{1}{6}x^2 = \dfrac{2}{6}x^2 - \dfrac{1}{6}x^2 + \dfrac{1}{4}x$ $\qquad\qquad\qquad = \dfrac{1}{6}x^2 + \dfrac{1}{4}x$	**6.** Combine like terms. **(a)** $-3a^2 + 6a + 4 + a^2 - 5a - 15$ **(b)** $5.2z^3 - 5.6z^2 + 4.9z^3 - 2.3z$
7. Use the distributive property to multiply. $-2x(x^2 + 5x)$ The distributive property tells us to multiply each term in the parentheses by the term outside the parentheses. $-2x(x^2 + 5x) = (-2x)(x^2) + (-2x)(5x)$ $\qquad\qquad\qquad = -2x^3 - 10x^2$	**8.** Use the distributive property to multiply. $-3a(a^3 + 4a^2)$

Copyright © 2013 Pearson Education, Inc.

Example	Student Practice
9. Multiply.	**10.** Multiply.

9. Multiply.

(a) $7x\left(x^2 - 3x - 5\right)$

Use the distributive property.

$7x\left(x^2 - 3x - 5\right) = 7x^3 - 21x^2 - 35x$

(b) $5ab\left(a^2 - ab + 8b^2 + 2\right)$

$5ab\left(a^2 - ab + 8b^2 + 2\right)$

$= 5a^3b - 5a^2b^2 + 40ab^3 + 10ab$

10. Multiply.

(a) $-4a\left(3a + a^2 + 2\right)$

(b) $6yz^2\left(-2y + 3z^2 - y^2 + 4\right)$

11. Simplify.

(a) $-(2x + 3)$

A parenthesis proceeded by a $(-)$ is assumed to have a coefficient of -1.

$-(2x + 3) = -1(2x + 3)$

$\qquad = -2x - 3$

(b) $\dfrac{2}{3}\left(6x^2 - 2x + 3\right)$

Use the distributive property.

$\dfrac{2}{3}\left(6x^2 - 2x + 3\right)$

$= \dfrac{2}{3}\left(6x^2\right) - \dfrac{2}{3}(2x) + \dfrac{2}{3}(3)$

$= 4x^2 - \dfrac{4}{3}x + 2$

12. Simplify.

(a) $-\left(2y^3 + 3y^2 - y + 7\right)$

(b) $-7a\left(-2a + 3b\right)$

Copyright © 2013 Pearson Education, Inc.

Example	Student Practice
13. Simplify. $5(x-2y)-(y+3x)+(5x-8y)$ $5(x-2y)-(y+3x)+(5x-8y)$ $=5(x-2y)-1(y+3x)+1(5x-8y)$ $=5x-10y-y-3x+5x-8y$ $=10x-19y$	**14.** Simplify. $-3(a+4b)-(2a-b)+(6a-7b)$
15. Simplify. $-2\{3+2[x-4(x+y)]\}$ $-2\{3+2[x-4(x+y)]\}$ $=-2\{3+2[x-4x-4y]\}$ $=-2\{3+2[-3x-4y]\}$ $=-2\{3-6x+8y\}$ $=-6+12x-16y$	**16.** Simplify. $-\{3a-2[b+3(a-2b)]\}$

Extra Practice

1. Combine like terms. $x^3-3x+x+2x^3$

2. Multiply. $-y(y^2-3y+1)$

3. Multiply. $\dfrac{x}{3}(2x^3-3x^2+9)$

4. Simplify. $-2+2(-2y^3+10)+2y(6y^2-1)$

Concept Check

Explain how to simplify the following. $2x^2-3x+4y-2x^2y-8x-5y$

Copyright © 2013 Pearson Education, Inc.

Name: _____   Date: _____
Instructor: _____   Section: _____

Chapter 1 Basic Concepts
1.6 Evaluating Variable Expressions and Formulas

Vocabulary
evaluate • formula • order of operations • perimeter • area • circumference

1. A(n) _____ is a rule for finding the value of a variable when the values of other variables in the expression are known.

2. When evaluating a variable expression, first replace each variable by its numerical value, then carry out each step, using the correct _____.

3. To _____ a variable expression, substitute the known values of the variables into the expression.

Example	Student Practice
1. Evaluate $x^2 - 5x - 6$ when $x = -4$.	**2.** Evaluate $x^2 + 2x - 3$ when $x = -5$.

Replace x by -4 and put parentheses around it. Then, simplify.

$$(-4)^2 - 5(-4) - 6 = 16 - 5(-4) - 6$$
$$= 16 - (-20) - 6$$
$$= 16 + 20 - 6$$
$$= 30$$

3. Evaluate $(5-x)^2 + 3xy$ when $x = -2$ and $y = 3$.

$$(5-x)^2 + 3xy = [5-(-2)]^2 + 3(-2)(3)$$
$$= [5+2]^2 + 3(-2)(3)$$
$$= 7^2 + 3(-2)(3)$$
$$= 49 + 3(-2)(3)$$
$$= 49 - 18 = 31$$

4. Evaluate $7xy - (4-x)^2$ when $x = 2$ and $y = -6$.

Vocabulary Answers: 1. formula 2. order of operations 3. evaluate

Copyright © 2013 Pearson Education, Inc.

Example	Student Practice
5. Evaluate when $x = -3$. (a) $(-2x)^2$ $(-2x)^2 = [-2(-3)]^2 = [6]^2 = 36$ (b) $-2x^2$ $-2x^2 = -2(-3)^2 = -2(9) = -18$	**6.** Evaluate when $x = -6$. (a) $(-4x)^2$ (b) $-4x^2$
7. Find the Fahrenheit temperature when the Celsius temperature is $-30°C$. Substitute the known value -30 for the variable C and evaluate. $F = \dfrac{9}{5}C + 32 = \dfrac{9}{5}(-30) + 32$ $= \dfrac{9}{\cancel{5}}\left(-\cancel{30}^{\,6}\right) + 32 = 9(-6) + 32$ $= -54 + 32 = -22$ Thus, the equivalent Fahrenheit temperature is $-22°F$.	**8.** Find the Fahrenheit temperature when the Celsius temperature is $-45°C$.
9. An amount of money invested or borrowed (not including interest) is called principal. Find the amount A to be repaid on a principal p of $1000 borrowed at a simple interest rate r of 8% for a time t of 2 years. The formula is $A = p(1 + rt)$. $A = p(1 + rt)$ $\quad = 1000[1 + (0.08)(2)]$ $\quad = 1000(1.16) = 1160$ The amount to be repaid A is $1160.	**10.** An amount of money invested or borrowed (not including interest) is called principal. Find the amount A to be repaid on a principal p of $2200 borrowed at a simple interest rate r of 3% for a time t of 5 years. The formula is $A = p(1 + rt)$.

Copyright © 2013 Pearson Education, Inc.

Example	Student Practice

11. Find the perimeter of a rectangular school playground with length 28 meters and width 16.5 meters. Use the formula $P = 2l + 2w$.

First, draw a picture to get a better idea of the situation. Then, substitute the known values 28 and 16.5 for the variables l and w, respectively, and simplify.

$P = 2l + 2w$

$\quad = 2(28) + 2(16.5)$

$\quad = 56 + 33$

$\quad = 89$

The perimeter of the playground is 89 meters.

12. Find the perimeter of a rectangular pool with length 18 feet and width 12 feet. Use the formula $P = 2l + 2w$.

---

13. Find the area of a trapezoid that has a height of 6 meters and bases of 7 meters and 11 meters.

First, draw a picture.

$b = 7$ meters

$a = 6$ meters

$c = 11$ meters

The formula is $A = \dfrac{1}{2} a(b + c)$.

Substitute the known values and simplify.

$A = \dfrac{1}{2} a(b + c) = \dfrac{1}{2}(6)(7 + 11)$

$\quad = \dfrac{1}{2}(6)(18)$

$\quad = 54$

The area is 54 square meters or 54 m^2.

14. Find the area of a triangle that has a height of 16 feet and base 10 feet in length. Use the formula $A = \dfrac{1}{2} ab$.

Copyright © 2013 Pearson Education, Inc.

Example	Student Practice
15. Find the volume of a sphere with a radius of 3 centimeters.	**16.** Find the volume of a sphere with a radius of 6 inches.

Use the formula $V = \dfrac{4}{3}\pi r^3$.

$$V = \frac{4}{3}\pi r^3 \approx \frac{4}{3}(3.14)(3)^3$$

$$= \frac{4}{3}(3.14)(27)$$

$$= \frac{4}{\overset{}{\cancel{3}}_1}(3.14)\left(\overset{9}{\cancel{27}}\right) = 113.04$$

The volume is approximately 113.04 cubic centimeters or 113.04 cm^3.

Extra Practice

1. Evaluate $8x^2 + x$ when $x = 4$.

2. Evaluate $\dfrac{3x + 5y + 2}{2x + y}$ when $x = 4$ and $y = 5$.

3. Find the amount to be repaid on a loan of $1800 at a simple interest rate of 12% for 2 years. Use the simple interest formula $A = p(1 + rt)$.

4. Find the circumference of a circle with a diameter of 9 inches. Use the formula $C = \pi d$.

Concept Check

Explain how you would find the amount A to be repaid on a principal of $5000 borrowed at a simple interest rate r of 8% for a time t of 2 years. The formula is $A = p(1 + rt)$.

Copyright © 2013 Pearson Education, Inc.

MATH COACH

Mastering the skills you need to do well on the test.

Watch the **MATH COACH** videos in MyMathLab® or on You Tube™
while you work the problems below. These helpful hints will
help you avoid making common errors on test problems.

Using the Order of Operations with Grouping Symbols and
Many Operations—Problem 4

Evaluate. $(7-5)^3 - 18 \div (-3) + \sqrt{10+6}$

Helpful Hint: First combine any numbers inside grouping symbols
(parentheses, radicals). Next, evaluate exponents and evaluate roots.

Did you first simplify the expression to
$(2)^3 - 18 \div (-3) + \sqrt{16}$? Yes _____ No _____

If you answered Problem 4 incorrectly, go
back and rework the problem using these
suggestions.

If you answered No, please go back and carefully combine
the numbers inside the grouping symbols.

Did you next simplify the expression to $8 - 18 \div (-3) + 4$?
Yes _____ No _____

If you answered No, please go back and raise the number 2
to the third power. Then find the square root of 16.

Using the Power Rule with Negative Exponents—Problem 8

Simplify. $\left(\dfrac{5a^{-2}b}{a} \right)^2$

Helpful Hint: First remove the parentheses by using the power rule for exponents. Then write the expression
using only positive exponents.

When you used the power rule for exponents, did you get
$\dfrac{5^2 a^{-4} b^2}{a^2}$? Yes _____ No _____

Now go back and rework the problem using
these suggestions.

If you answered No, remember that the power rule allows
you to multiply the exponents.

When you used the rule of negative exponents, did you get
a^6 as your denominator? Yes _____ No _____

If you answered No, remember that $a^4 \left(a^2 \right)$ requires you to
use the product rule of exponents.

Copyright © 2013 Pearson Education, Inc.

Simplifying Algebraic Expressions with Many Grouping Symbols—Problem 16

Simplify. $2\left[-3(2x+4)+8(3x-2)\right]$

Helpful Hint: Work from the inside out by removing the innermost parentheses first. Then work within the resulting expression inside the brackets.

Did you first apply the distributive property to remove the innermost parentheses and obtain the expression
$2\left[-6x-12+24x-16\right]$?

If you answered Problem 16 incorrectly, go back and rework the problem using these suggestions.

Yes _____ No _____

If you answered No, stop and perform that step again. Be careful to avoid sign errors.

Were you able to combine like terms to get $2\left[18x-28\right]$?

Yes _____ No _____

If you answered No, go back and take the time to combine the x terms and then combine the constant terms. Then finish the problem on your own.

Evaluating a Variable Expression Given the Values of the Variables—Problem 18

Evaluate $5x^2+3xy-y^2$ when $x=3$ and $y=-3$.

Helpful Hint: First, replace each variable with its given value. Place parentheses around each numerical value when you make the substitution. Then use the proper order of operations.

In your first step, did you write $5(3)^2+3(3)(-3)-(-3)^2$?

Now go back and rework the problem using these suggestions.

Yes _____ No _____

If you answered No, stop and carefully make this substitution. Be sure to surround the number you are substituting with parentheses.

After evaluating the exponents, did you get the expression $5(9)+3(3)(-3)-(9)$?

Yes _____ No _____

If you answered No, go back and square 3 and then square -3. Remember that $(-3)^2=9$.

26

Copyright © 2013 Pearson Education, Inc.

Chapter 2 Linear Equations and Inequalities
2.1 First-Degree Equations with One Unknown

Vocabulary

equation • first-degree equation with one unknown • solve • solution
linear equation with one unknown • root • equivalent • no solution
any real number is a solution • terminating decimal

1. To _____ a first-degree equation with one unknown, find the value of the variable that makes the equation a true mathematical statement.

2. Equations that have the same solution are said to be _____.

3. Another phrase for a first-degree equation with one unknown is _____.

4. Another word for the solution of an equation is _____.

Example	Student Practice
1. Is $\dfrac{1}{3}$ a solution of the equation $2a+5=a+6$?	**2.** Is $\dfrac{1}{4}$ a solution of the equation $4x-3=-12x+1$?

We replace a by $\dfrac{1}{3}$ in the equation $2a+5=a+6$.

$$2a+5=a+6$$

$$2\left(\frac{1}{3}\right)+5\overset{?}{=}\frac{1}{3}+6$$

$$\frac{2}{3}+5\overset{?}{=}\frac{1}{3}+6$$

$$\frac{17}{3}\neq\frac{19}{3}$$

This last statement is not true. Thus, $\dfrac{1}{3}$ is not a solution of $2a+5=a+6$.

Vocabulary Answers: 1. solve 2. equivalent 3. linear equation with one unknown 4. root

Copyright © 2013 Pearson Education, Inc.

Example	Student Practice
3. Solve. $\dfrac{1}{3}y = -6$	**4.** Solve. $\dfrac{1}{4}x = -6$

Multiply each side by 3 to eliminate the fraction.

$$3\left(\dfrac{1}{3}\right)y = 3(-6)$$

$$y = -18$$

The solution is -18. The check is left to the student.

---

5. Solve. $6x - 2 - 4x = 8x + 3$

$$6x - 2 - 4x = 8x + 3$$
$$2x - 2 = 8x + 3$$
$$2x - 8x - 2 = 8x - 8x + 3$$
$$-6x - 2 = 3$$
$$-6x - 2 + 2 = 3 + 2$$
$$-6x = 5$$
$$\dfrac{-6x}{-6} = \dfrac{5}{-6}$$
$$x = -\dfrac{5}{6}$$

When checking fractional values like $x = -\dfrac{5}{6}$, take extra care in order to perform the operations correctly.

$$6\left(-\dfrac{5}{6}\right) - 2 - 4\left(-\dfrac{5}{6}\right) \overset{?}{=} 8\left(-\dfrac{5}{6}\right) + 3$$

$$-5 - 2 + \dfrac{10}{3} \overset{?}{=} \dfrac{-20}{3} + 3$$

$$\dfrac{-21}{3} + \dfrac{10}{3} \overset{?}{=} \dfrac{-20}{3} + \dfrac{9}{3}$$

$$-\dfrac{11}{3} = -\dfrac{11}{3}$$

6. Solve. $5x + 6 = 4x + 8x - 7 + 3$

Copyright © 2013 Pearson Education, Inc.

Example	Student Practice
7. Solve. $\dfrac{x}{5}+\dfrac{1}{2}=\dfrac{4}{5}+\dfrac{x}{2}$	**8.** Solve. $\dfrac{x}{6}+\dfrac{1}{3}=\dfrac{x}{2}+\dfrac{5}{3}$

$$\frac{x}{5}+\frac{1}{2}=\frac{4}{5}+\frac{x}{2}$$

$$10\left(\frac{x}{5}+\frac{1}{2}\right)=10\left(\frac{4}{5}+\frac{x}{2}\right)$$

$$10\left(\frac{x}{5}\right)+10\left(\frac{1}{2}\right)=10\left(\frac{4}{5}\right)+10\left(\frac{x}{2}\right)$$

$$2x+5=8+5x$$

$$2x-2x+5=8+5x-2x$$

$$5=8+3x$$

$$5-8=8-8+3x$$

$$-3=3x$$

$$\frac{-3}{3}=\frac{3x}{3}$$

$$-1=x$$

The check is left to the student.

9. Solve and check.	**10.** Solve and check.
$0.9+0.2(x+4)=-3(0.1x-0.4)$	$0.5x+6(0.1x-0.4)=-0.4(x-9)$

$$0.9+0.2(x+4)=-3(0.1x-0.4)$$

$$0.9+0.2x+0.8=-0.3x+1.2$$

$$0.2x+1.7=-0.3x+1.2$$

$$10(0.2x)+10(1.7)=10(-0.3x)+10(1.2)$$

$$2x+17=-3x+12$$

$$2x+3x+17=-3x+3x+12$$

$$5x+17=12$$

$$5x+17-17=12-17$$

$$5x=-5$$

$$x=-1$$

The check is left to the student.

Copyright © 2013 Pearson Education, Inc.

Example	**Student Practice**
11. Solve. $7x+3-9x=14-2x+5$	**12.** Solve. $23+6x-13=14x+6-8x$

$$7x+3-9x=14-2x+5$$
$$-2x+3=-2x+19$$
$$-2x+3+2x=-2x+19+2x$$
$$3=19$$

No matter what value we use for x in this equation, we get a false sentence. This equation has no solution.

13. Solve. $5(x-2)+3x=10x-2(x+5)$	**14.** Solve. $12x+4(x+3)=8x+25+8x-13$

$$5(x-2)+3x=10x-2(x+5)$$
$$5x-10+3x=10x-2x-10$$
$$8x-10=8x-10$$
$$8x-10-8x=8x-10-8x$$
$$-10=-10$$

Replacing x in this equation by any real number will always result in a true sentence. This is an equation for which any real number is a solution.

Extra Practice

1. Solve. $4x-3=13$

2. Solve. $(y-5)-(y+8)=9y$

3. Solve. $\dfrac{4x}{5}-\dfrac{2x}{3}=6$

4. Solve. $2.6x+4=0.2(3x+2)+1.6$

Concept Check

Explain how you would solve the following equation for x. $\dfrac{3x+1}{2}+\dfrac{2}{3}=\dfrac{4x}{5}$

Copyright © 2013 Pearson Education, Inc.

Chapter 2 Linear Equations and Inequalities
2.2 Literal Equations and Formulas

Vocabulary
literal equation   •   grouping symbols   •   LCD

1. The first step in solving linear literal equations is to remove _____.

2. A first-degree _____ is an equation that contains at least one variable other than the variable that we are solving for.

3. If a literal equation contains fractions, multiply each term by the _____.

Example	Student Practice
1. Solve for x. $5x + 3y = 2$	**2.** Solve for y. $5x + 3y = 2$

$$5x + 3y = 2$$
$$5x = 2 - 3y$$
$$\frac{5x}{5} = \frac{2-3y}{5}$$
$$x = \frac{2-3y}{5}$$

3. Solve for b. $A = \frac{2}{3}(a + b + 3)$ | **4.** Solve for z. $A = \frac{4}{5}(x - y + 2z)$

$$A = \frac{2}{3}(a + b + 3)$$
$$A = \frac{2}{3}a + \frac{2}{3}b + 2$$
$$3A = 3\left(\frac{2}{3}a\right) + 3\left(\frac{2}{3}b\right) + 3(2)$$
$$3A = 2a + 2b + 6$$
$$3A - 2a - 6 = 2b$$
$$\frac{3A - 2a - 6}{2} = \frac{2b}{2}$$
$$\frac{3A - 2a - 6}{2} = b$$

Vocabulary Answers: 1. grouping symbols 2. literal equation 3. LCD

Copyright © 2013 Pearson Education, Inc.

Example	Student Practice
5. Solve for x. $5(2ax+3y)-4ax=2(ax-5)$	**6.** Solve for z. $-3a(4z+4x)-2(5a-4z)=-7ax$

5. Solve for x.

$$5(2ax+3y)-4ax=2(ax-5)$$
$$10ax+15y-4ax=2ax-10$$
$$6ax-2ax+15y=-10$$
$$4ax=-10-15y$$
$$\frac{4ax}{4a}=\frac{-10-15y}{4a}$$
$$x=\frac{-10-15y}{4a}$$

---

7. The world record times in minutes for the men's marathon are approximated by the equation $t=-0.2x+135$, where x is the number of years since 1960. Solve this equation for x. Determine what year it will be when the world record time for the men's marathon is 122 minutes.

$$t=-0.2x+135$$
$$0.2x=135-t$$
$$2x=1350-10t$$
$$x=\frac{1350-10t}{2}$$

Now use this equation to find when the winning time is 122 minutes.

$$x=\frac{1350-10(122)}{2}$$
$$x=\frac{1350-1220}{2}$$
$$x=\frac{130}{2}=65$$

It will be 65 years from 1960. Thus, this winning time will occur in 2025.

8. The winning time for a local triathlon can be approximated by the equation $t=-0.3x+97$, where x is the number of years since 1985. Solve this equation for x. Determine approximately what year it will be when the winning time for this triathlon is 88 minutes.

Copyright © 2013 Pearson Education, Inc.

Example	Student Practice
9. Answer parts **(a)** and **(b)**.	**10.** Answer parts **(a)** and **(b)**.

(a) Solve the formula for the area of a trapezoid, $A = \dfrac{1}{2}a(b+c)$, for c.

(a) Solve for n. $A = a \cdot r(n-1)$

Remove the parentheses, clear the fractions, then solve for c.

$$A = \frac{1}{2}a(b+c)$$

$$A = \frac{1}{2}ab + \frac{1}{2}ac$$

$$2A = 2\left(\frac{1}{2}ab\right) + 2\left(\frac{1}{2}ac\right)$$

$$2A = ab + ac$$

$$2A - ab = ac$$

$$\frac{2A - ab}{a} = c$$

(b) Find c when $A = 20$ square inches, $a = 3$ inches, and $b = 4$ inches.

(b) Find n when $A = 50$, $a = 8$, and $r = 5$.

We use the equation we derived in part **(a)** to find c for the given values.

$$c = \frac{2A - ab}{a}$$

$$= \frac{2(20) - (3)(4)}{3}$$

$$= \frac{40 - 12}{3}$$

$$= \frac{28}{3}$$

Thus, side $c = \dfrac{28}{3}$ inches or $9\dfrac{1}{3}$ inches.

Copyright © 2013 Pearson Education, Inc.

Extra Practice

1. Solve for x. $y = \dfrac{5}{7}x - 15$

2. Solve for t. $I = prt$

3. Solve for y. $w = \dfrac{2y - x}{y}$

4. The formula for the surface area of a rectangular prism is $S = 2lw + 2lh + 2wh$, where S is the surface area of the prism, l is its length, w is its width, and h is its height.

    (a) Solve for the variable h.

    (b) Use this result to find the height of a rectangular prism with length 5 inches, width 6 inches, and surface area 126 square inches.

Concept Check

Explain how you would solve for x in the following equation. $3(3ax + y) = 2ax - 5y$

Copyright © 2013 Pearson Education, Inc.

Chapter 2 Linear Equations and Inequalities
2.3 Absolute Value Equations

Vocabulary
absolute value   • $|a| = |b|$

1. If _____, then $a = b$ or $a = -b$.

2. The _____ of a number x can be pictured as the distance between 0 and x on the number line.

Example	Student Practice
1. Solve and check your solutions. $$|2x + 5| = 11$$	**2.** Solve and check your solutions. $$|4x + 6| = 18$$

Example

1. Solve and check your solutions.
$$|2x + 5| = 11$$

The solutions of an equation of the form $|ax + b| = c$, where $a \neq 0$ and c is a positive number, are those values that satisfy $ax + b = c$ or $ax + b = -c$. Thus, we have the following:

$$2x + 5 = 11 \quad \text{or} \quad 2x + 5 = -11$$
$$2x = 6 \qquad\qquad 2x = -16$$
$$x = 3 \qquad\qquad\quad x = -8$$

The two solutions are 3 and −8. Check the solutions.

$$|2x + 5| = 11 \qquad |2x + 5| = 11$$
$$|2(3) + 5| \overset{?}{=} 11 \qquad |2(-8) + 5| \overset{?}{=} 11$$
$$|6 + 5| \overset{?}{=} 11 \qquad |-16 + 5| \overset{?}{=} 11$$
$$|11| \overset{?}{=} 11 \qquad |-11| \overset{?}{=} 11$$
$$11 = 11 \qquad\qquad 11 = 11$$

The solutions check.

Student Practice

2. Solve and check your solutions.
$$|4x + 6| = 18$$

Vocabulary Answers: 1. $|a| = |b|$ 2. absolute value

Copyright © 2013 Pearson Education, Inc.

Example	Student Practice
3. Solve and check your solutions. $$\left|\frac{1}{2}x - 1\right| = 5$$ The solutions of the given absolute value equation must satisfy $\frac{1}{2}x - 1 = 5$ or $\frac{1}{2}x - 1 = -5$. If we multiply each term of both equations by 2, we obtain the following: $x - 2 = 10$ or $x - 2 = -10$ $\quad\quad x = 12 \quad\quad\quad\quad x = -8$ The check is left to the student.	**4.** Solve and check your solutions. $$\left|\frac{2}{5}x + 6\right| = 4$$
5. Solve $|3x - 1| + 2 = 5$ and check your solutions. First we will rewrite the equation so that the absolute value expression is alone on one side of the equation. $$|3x - 1| + 2 = 5$$ $$|3x - 1| + 2 - 2 = 5 - 2$$ $$|3x - 1| = 3$$ Now we solve $|3x - 1| = 3$. $3x - 1 = 3$ or $3x - 1 = -3$ $\quad 3x = 4 \quad\quad\quad\quad 3x = -2$ $\quad\quad x = \dfrac{4}{3} \quad\quad\quad\quad x = -\dfrac{2}{3}$ The check is left to the student.	**6.** Solve $|6x - 3| - 4 = 5$ and check your solutions.

Copyright © 2013 Pearson Education, Inc.

Example	Student Practice
7. Solve and check. $\lvert 3x - 4 \rvert = \lvert x + 6 \rvert$	**8.** Solve and check. $\lvert x + 4 \rvert = \lvert 4x + 6 \rvert$

The solutions of the given equation must satisfy $3x - 4 = x + 6$ or $3x - 4 = -(x + 6)$. Now we solve each equation in the normal fashion.

$$3x - 4 = x + 6 \quad \text{or} \quad 3x - 4 = -x - 6$$
$$3x - x = 4 + 6 \qquad 3x + x = 4 - 6$$
$$2x = 10 \qquad\qquad 4x = -2$$
$$x = 5 \qquad\qquad\quad x = -\frac{1}{2}$$

We will check each solution by substituting it into the original equation.

if $x = 5$

$$\lvert 3(5) - 4 \rvert \overset{?}{=} \lvert 5 + 6 \rvert$$
$$\lvert 15 - 4 \rvert \overset{?}{=} \lvert 11 \rvert$$
$$\lvert 11 \rvert \overset{?}{=} \lvert 11 \rvert$$
$$11 = 11$$

if $x = -\frac{1}{2}$

$$\left\lvert 3\left(-\frac{1}{2}\right) - 4 \right\rvert \overset{?}{=} \left\lvert -\frac{1}{2} + 6 \right\rvert$$
$$\left\lvert -\frac{3}{2} - 4 \right\rvert \overset{?}{=} \left\lvert -\frac{1}{2} + 6 \right\rvert$$
$$\left\lvert -\frac{3}{2} - \frac{8}{2} \right\rvert \overset{?}{=} \left\lvert -\frac{1}{2} + \frac{12}{2} \right\rvert$$
$$\left\lvert -\frac{11}{2} \right\rvert \overset{?}{=} \left\lvert \frac{11}{2} \right\rvert$$
$$\frac{11}{2} = \frac{11}{2}$$

Copyright © 2013 Pearson Education, Inc.

Extra Practice

1. Solve. Check your solutions. $|2x+9|=31$

2. Solve. Check your solutions.
$$|3m+2|-10=-6$$

3. Solve. Check your solutions.
$$|4s+9|=|s+2|$$

4. Solve. Check your solutions.
$$\left|\frac{4x+3}{5}\right|=|3x+7|$$

Concept Check

Explain how you would solve for x in the following equation. $|2x+4|=\dfrac{1}{2}$

Copyright © 2013 Pearson Education, Inc.

Chapter 2 Linear Equations and Inequalities
2.4 Using Equations to Solve Word Problems

Vocabulary

more than   •   decreased by   •   product of   •   ratio of   •   will be

1. The phrase _____ usually denotes division.

2. The phrase _____ usually denotes addition.

3. The phrase _____ usually denotes equals.

4. The phrase _____ usually denotes subtraction.

Example	Student Practice
1. Nancy went to a local truck rental company to rent a truck. The truck rental company has a fixed rate of $40 per day plus 20¢ per mile. Nancy rented a truck for 3 days and was billed $117. How many miles did she drive?	**2.** A cell phone plan charges a flat monthly fee of $25 for the first 500 minutes, then 8¢ for each additional minute. A teenager's cell phone bill for last month came to $67. How many additional minutes did the teenager use?

Let n = the number of miles driven. Since each mile costs 20¢, we multiply the 20¢ (or $0.20) per-mile cost by the number of miles n. Thus, $0.20n$ = the cost of driving n miles at 20¢ per mile.

Write an equation and then solve and state the answer. Fixed costs for 3 days plus mileage charge equals $177.

$$(40)(3)+(0.20)(n)=177$$
$$120+0.20n=177$$
$$0.20n=57$$
$$\frac{0.20n}{0.20}=\frac{57}{0.20}$$
$$n=285$$

The check is left to the student.

Vocabulary Answers: 1. ratio of 2. more than 3. will be 4. decreased by

Copyright © 2013 Pearson Education, Inc.

Example	Student Practice

3. The Acetones are a barbershop quartet. They travel across the country in a special bus and usually give six concerts a week. This popular group always sings to a sell-out crowd. The concert halls have an average seating capacity of three thousand people each. Concert tickets average $12 per person. The onetime expenses for each concert are $15,000. The cost of meals, motels, security and sound people, bus drivers, and other expenses totals $100,000 per week. How many weeks per year will the Acetones need to be on tour if each member wants to earn $71,500 per year?

The item we are trying to find is the number of weeks the quartet needs to be on tour. So we let $x =$ the number of weeks on tour. Now we need to find an expression that describes the quartet's income. Each week, there are six concerts with three thousand people paying $12 per person, Thus, weekly income $= (6)(3000)(12) = \$216,000$.

and income for x weeks $= \$216,000x$. Now we need to find an expression for the total expenses. The expenses for each concert are $15,000, and there are six concerts per week. Thus, concert expenses will be $(6)(15,000) = \$90,000$

per week. Now the cost per week totals $100,000. Thus, total weekly expenses $= \$90,000 + \$100,000 = 190,000,$ and total expenses for x weeks $= \$190,000x$. The four Acetones each want to earn $71,500 per year, so the group will need $4 \times \$71,500 = \$286,000$ for the year.

$$\$216,000x - 190,000x = \$286,000$$
$$x = 11$$

4. A country music band, consisting of 3 members, is going on tour. They are scheduled to give 6 concerts a week with an average audience of seven thousand people at each concert. The average ticket price for a concert is $10. The onetime expenses for each concert are $44,000. The additional costs per week are $135,000. How many weeks per year will this band need to be on tour if each member wants to earn $210,000 per year?

Copyright © 2013 Pearson Education, Inc.

Example	Student Practice

5. An astronaut's space suit contains a small rectangular steel plate that supports the breathing control valve. The length of the rectangle is 3 millimeters more than double its width. Its perimeter is 108 millimeters. Find the width and length.

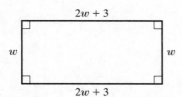

The formula for the perimeter of a rectangle is $P = 2w + 2l$, where $w =$ the width and $l =$ the length. Since the length is compared to the width, let $w =$ the width. Then $2w + 3 =$ the length.

$P = 2w + 2l$

$108 = 2w + 2(2w + 3)$

$108 = 2w + 4w + 6$

$108 = 6w + 6$

$102 = 6w$

$17 = w$

Because $w = 17$, we have $2w + 3 = 2(17) + 3 = 37$. Thus the rectangle is 17 millimeters wide and 37 millimeters long.

Check.

$P = 2w + 2l$

$108 \overset{?}{=} 2(17) + 2(37)$

$108 \overset{?}{=} 34 + 74$

$108 = 108$

6. The perimeter of a rectangle is 154 yards. The length is 28 yards less than double the width. Find each dimension.

Copyright © 2013 Pearson Education, Inc.

1. The sum of 3 and the quotient of a number and 4 is 22. What is the number?

2. Damian is reading a novel. Yesterday, Damian read 30 fewer than twice the number of pages he read today. Today, he read 84 pages. How many pages did he read yesterday?

3. Simon can rent a television for $95 per month, or he can buy the television for $855. In how many months will the cost of renting the television equal the cost of buying the television?

4. In October, Natasha, Mia, and Frederick volunteered a total of 38 hours at a local soup kitchen. Natasha worked 6 hours more than Frederick, and Mia worked twice as many hours as Frederick. How many hours did each of them work?

Concept Check

Explain how you would set up an equation to solve the following problem, then solve the problem. A driveway has a perimeter of 212 feet. The length of the driveway is 7 feet longer than four times the width. Find the width and length of the driveway.

Copyright © 2013 Pearson Education, Inc.

Chapter 2 Linear Equations and Inequalities
2.5 Solving More-Involved Word Problems

Vocabulary
simple interest   •   mixture problem

1. A problem in which two or more items are combined to form a mixture or solution is called a _____.

2. _____ is an income from investing money or a charge for borrowing money.

Example	Student Practice
1. The Wildlife Refuge Rangers tagged 144 deer. They estimate that they have tagged 36% of the deer in the refuge. If they are correct, approximately how many deer are in the refuge?	**2.** In a recent election of a small town, 225 of the residents voted. They estimate that 20% of the townspeople voted in the election. Approximately how many townspeople are there in the town?

Let $n =$ the number of the deer in the refuge. Then $0.36n = 36\%$ of the deer in the refuge.

36% of the deer in the refuge gives a total of 144 tagged deer.
$0.36 \times n = 144$

$$0.36n = 144$$
$$\frac{0.36n}{0.36} = \frac{144}{0.36}$$
$$n = 400$$

There are approximately 400 deer in the refuge.

Check. Is it true that 36% of 400 is 144?

$$(0.36)(400)\overset{?}{=}144$$
$$144 = 144$$

It checks. Our answer is correct.

Vocabulary Answers: 1. mixture problem 2. simple interest

Example	Student Practice

3. Maria has a job as a financial advisor in a bank. She advised a customer to invest part of his money in a money market fund earning 12% simple interest and the rest in an investment fund earning 14% simple interest. The customer had $6000 to invest. If he earned $772 in interest in 1 year, how much did he invest in each fund?

Let x = the amount of money invested at 12% interest. The other amount of money is (total $-x$).
Thus, $6000 - x$ = the amount of money invested at 14% interest.

Write an equation. Then solve and state the answer.

Interest earned at 12%, or 12% of x, added to interest earned at 14%, or 14% of $6000 - x$, is equal to total interest earnings of $772.

$$0.12x + 0.14(6000 - x) = 772$$
$$0.12x + 840 - 0.14x = 772$$
$$840 - 0.02x = 772$$
$$-0.02x = -68$$
$$\frac{-0.02x}{-0.02} = \frac{-68}{-0.02}$$
$$x = 3400$$

If $x = 3400$, then $6000 - x$ $= 6000 - 3400 = 2600$. Thus, $3400 was invested in the money market fund earning 12% interest, and $2600 was invested in the investment fund earning 14% interest.

The check is left to the student.

4. Gary won $6000 in a local lottery. He invested part of it at 11% simple interest and the remainder at 16% simple interest. At the end of the year he had earned $835 in interest. How much did Gary invest at each interest amount?

Copyright © 2013 Pearson Education, Inc.

Example	Student Practice

5. A small truck has a radiator that holds 20 liters. A mechanic needs to fill the radiator with a solution that is 60% antifreeze. He has 70% and 30% antifreeze solutions. How many liters of each should he use to achieve the desired mix?

Let x = the number of liters of 70% antifreeze to be used. Since the total amount of solution must be 20 liters, we can use $20 - x$ for the other part. So $20 - x$ = the number of liters of 30% antifreeze to be used.

Write an equation. Then solve and state the answer.

Number of liters of 70% antifreeze added to number of liters of 30% antifreeze gives a resulting solution of 20 liters of 60% antifreeze.

$$0.70x + 0.30(20 - x) = 0.60(20)$$
$$0.70x + 6 - 0.30x = 12$$
$$6 + 0.40x = 12$$
$$0.40x = 6$$
$$\frac{0.40x}{0.40} = \frac{6}{0.40}$$
$$x = 15$$

If $x = 15$, then $20 - x = 20 - 15 = 5$. Thus, the mechanic needs 15 liters of 70% antifreeze solution and 5 liters of 30% antifreeze solution.

The check is left to the student.

6. A chef is mixing nuts for a mixed nut bag. How much $2 a pound nuts and $5 a pound nuts should be mixed together to get a 10 pound bag of $3.50 a pound mixed nuts.

Copyright © 2013 Pearson Education, Inc.

Example	Student Practice

7. Frank drove at a steady speed for 3 hours on the turnpike. He then slowed his speed by 15 miles per hour on the secondary roads. The entire trip took 5 hours and covered 245 miles. What was his speed on each portion of the trip?

Let $x =$ turnpike speed.
So $x - 15 =$ secondary roads speed.

Distance on turnpike + Distance on secondary roads = Total distance
$$3x + 2(x - 15) = 245$$
$$x = 55$$

Thus, Frank traveled 55 miles per hour on the turnpike and 40 miles per hour on the secondary road.

8. Fey drove for 3 hours at a steady speed. She increased her speed by 15 miles per hour for the last part of the trip. The entire trip took 4 hours and covered 215 miles. How fast did she drive on each portion of the trip?

Extra Practice

1. The apple tree in Raul's back yard grew 36 apples this year. This is an increase of 20% from the number of apples it grew last year. How many apples did it grow last year?

2. Dexter has $5000 to invest. He wants to invest some of the money in a savings account that earns 3% simple interest and the rest in a bond that earns 5% simple interest. If he wants to make a total of $220 in interest in one year, how much should he invest in the bond?

3. A jeweler wishes to make 150 grams of 70% pure gold from sources that are 80% pure gold and 55% pure gold. How many grams of each should he use?

4. Lamar drove at a constant speed for 3 hours in heavy traffic. The traffic thinned out and he increased his speed by 30 miles per hour for the last 2 hours of his 270-mile trip. What was his speed in heavy traffic and what was his speed when the traffic thinned out?

Concept Check
How would you set up an equation to solve the following problem? Then solve the problem. A new sports car sold for a certain amount of money. This year the price went up 12%. The new price is $39,200. What was the price the previous year?

Copyright © 2013 Pearson Education, Inc.

Chapter 2 Linear Equations and Inequalities
2.6 Linear Inequalities

Vocabulary
linear inequality   •   less than or equal to   •   greater than or equal to   •   equivalent

1. The symbol ≥ means _____.

2. The symbol ≤ means _____.

3. Inequalities that have the same solution are said to be _____.

4. A(n) _____ is a statement that describes how two numbers or linear expressions are related to one another.

Example	**Student Practice**
1. Insert the proper symbol between the numbers.	**2.** Insert the proper symbol between the numbers.
(a) $\dfrac{1}{2}$ _____ $\dfrac{1}{3}$	**(a)** $\dfrac{3}{7}$ _____ $\dfrac{2}{5}$
When comparing two fractions, rewrite them with a common denominator.	
$\dfrac{3}{6}$ _____ $\dfrac{2}{6}$ $\dfrac{3}{6} > \dfrac{2}{6}$	
Thus, $\dfrac{1}{2} > \dfrac{1}{3}$ because $\dfrac{3}{6} > \dfrac{2}{6}$.	**(b)** -0.423 _____ -0.44
(b) -0.033 _____ -0.0329	
$-0.033 < -0.0329$ because $-0.0330 < -0.0329$. Since both numbers are negative, -0.0330 is to the left of -0.0329 on a number line.	

Vocabulary Answers: 1. greater than or equal to 2. less than or equal to 3. equivalent 4. linear inequality

Copyright © 2013 Pearson Education, Inc.

Example	Student Practice
3. Insert the proper symbol between the expressions. $\left|1-7\right|\underline{\quad}\left|-4-12\right|$ Evaluate each expression and compare. $\left|1-7\right|<\left|-4-12\right|$ because $6<16$.	**4.** Insert the proper symbol between the expressions. $\left|5-13\right|\underline{\quad}\left|2-6\right|$

5. Graph each inequality.

  (a) $x<0$

  (b) $x\geq-5$

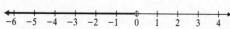

  (c) $2<x$

  We read an inequality starting with the variable. Thus, we read $2<x$ as "x is greater than 2." Graph the expression accordingly.

6. Graph each inequality.

  (a) $x<-2$

  (b) $x\leq-2$

  (c) $4>x$

---

7. Solve the inequality. Graph and check the solution. $x-8<15$

$$x-8<15$$
$$x-8+8<15+8$$
$$x<23$$

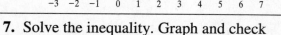

To check, we will choose 22.5.

$$x-8<15$$
$$22.5-8\overset{?}{<}15$$
$$14.5<15$$

8. Solve the inequality. Graph and check the solution. $x-5\leq14$

Copyright © 2013 Pearson Education, Inc.

Example	Student Practice
9. Solve and graph your solution. $6x + 3 \leq 2x - 5$ $6x + 3 \leq 2x - 5$ $6x \leq 2x - 8$ $4x \leq -8$ $x \leq -2$ The check is left to the student.	**10.** Solve and graph your solution. $9x + 4 \geq 5x - 8$
11. Solve. $-0.3x + 1.0 \leq 1.2x - 3.5$ $10(-0.3x + 1.0) \leq 10(1.2x - 3.5)$ $-3x + 10 \leq 12x - 35$ $-3x - 12x + 10 \leq 12x - 12x - 35$ $-15x + 10 \leq -35$ $-15x + 10 - 10 \leq -35 - 10$ $-15x \leq -45$ $\dfrac{-15x}{-15} \geq \dfrac{-45}{-15}$ $x \geq 3$	**12.** Solve. $-0.4x + 1.2 < 0.3x - 0.2$
13. Solve. $\dfrac{1}{7}(x + 5) > \dfrac{1}{5}(x + 1)$ $\dfrac{1}{7}(x + 5) > \dfrac{1}{5}(x + 1)$ $\dfrac{x}{7} + \dfrac{5}{7} > \dfrac{x}{5} + \dfrac{1}{5}$ $35\left(\dfrac{x}{7}\right) + 35\left(\dfrac{5}{7}\right) > 35\left(\dfrac{x}{5}\right) + 35\left(\dfrac{1}{5}\right)$ $5x + 25 > 7x + 7$ $5x > 7x - 18$ $-2x > -18$ $x < 9$	**14.** Solve. $\dfrac{1}{2}(x - 5) \leq \dfrac{1}{3}(x + 2)$

Copyright © 2013 Pearson Education, Inc.

Example	Student Practice

15. Lexi and her mother are using a public phone to make a long distance phone call from Honolulu, HI, to West Chicago, IL. The charge is $4.50 for the first minute and 85¢ for each additional minute. Any fractional part of a minute will be rounded up to the nearest whole minute. What is the maximum time that Lexi and her mother can talk if they have $15.55 in change to make the call?

Let x = the number of minutes they talk after the first minute. The cost must be less than or equal to $15.55.
$$4.50 + 0.85x \le 15.55$$
$$0.85x \le 11.05$$
$$x \le 13$$

Add 13 minutes to the one minute that cost $4.50. This gives us 14 minutes.

16. Two people are making a long distance phone call. The charge is $5.50 for the first minute and 42¢ for each additional minute. Any fractional part of a minute will be rounded up to the nearest whole minute. What is the maximum time that the two people can talk if they have $11.38 in change to make the call?

Extra Practice

1. Insert the symbol < or > between the pair of numbers. $-\dfrac{7}{12}$ —— $-\dfrac{5}{8}$

2. Solve for x and graph your solution.
$$x + 1 < -11$$

3. Solve for x and graph your solution.
$$0.8 + x \ge 0.9x + 0.4$$

4. Describe the situation with a linear inequality and then solve the inequality. A waiter earns $4 per hour plus an average tip of $7 for every table served. How many tables must he serve to earn more than $129 for a 6-hour shift?

Concept Check

If you were to solve the inequalities $-3x < 9$ and $3x < 9$, in one case you would have to reverse the direction of the inequality and in the other case you would not. Explain how you can tell which case is which.

Copyright © 2013 Pearson Education, Inc.

Name: _____   Date: _____
Instructor: _____   Section: _____

Chapter 2 Linear Equations and Inequalities
2.7 Compound Inequalities

Vocabulary
compound inequalities • empty set • and • or

1. The solution of a compound inequality using the connective _____ includes all the numbers that are solutions of either of the two inequalities.

2. Inequalities that consist of two inequalities connected by the word and or the word or are called _____.

3. The notation $\varnothing$ represents the _____.

4. The solution of a compound inequality using the connective _____ includes all the numbers that make both parts true at the same time.

Example	Student Practice
1. Graph the values of x where $7 < x$ and $x < 12$.	**2.** Graph the values of x where $-2 < x$ and $x < 2$.

We read the inequality starting with the variable. Thus, we graph all values of x, where x is greater than 7 and where x is less than 12. All such values must be between 7 and 12. Numbers that are greater than 7 and less than 12 can be written as $7 < x < 12$.

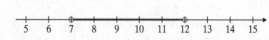

3. Graph the values of x where $-8.5 \leq x < -1$.

Note the shaded circle at -8.5 and the open circle at -1.

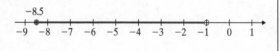

4. Graph the values of x where $4.5 \leq x \leq 6$

Vocabulary Answers: 1. or 2. compound inequalities 3. empty set 4. and

Copyright © 2013 Pearson Education, Inc.

Example	Student Practice
5. Graph the salary range (s) of the full-time employees of Tentron Corporation. Each person earns at least \$190 weekly, but not more than \$800 weekly.	**6.** Graph the salary range of a person who earns at least \$300 per week, but never more than \$1050 per week.

"At least \$190" means that the weekly salary of each person is greater than or equal to \$190. We write $s \geq \$190$. "Not more than" means that the weekly salary of each person is less than or equal to \$800. We write $s \leq \$800$. Thus, s may be between 190 and 800 and may include those endpoints.

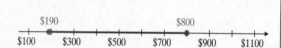

Example	Student Practice
7. Graph the region where $x < 3$ or $x > 6$.	**8.** Graph the region where $x < -1$ or $x > 1$.

Read the inequality as " x is less than 3 or x is greater than 6." This includes all values to the left of 3 as well as all values to the right of 6 on a number line. We shade these regions.

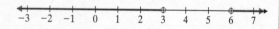

Example	Student Practice
9. Male applicants for the state police force in Fred's home state are ineligible for the force if they are shorter than 60 inches or taller than 76 inches. Graph the range of rejected applicants' heights.	**10.** Any applicant is ineligible if they are younger than 21 years old or older than 60 years old. Graph the range of rejected applicants' ages.

Each rejected applicant's height h will be less than 60 inches, $h < 60$ or will be greater than 76 inches, $h > 76$.

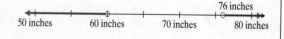

Copyright © 2013 Pearson Education, Inc.

Example	Student Practice
11. Solve for x and graph the compound solution. $3x+2>14$ or $2x-1<-7$	**12.** Solve for x and graph the compound solution. $4x-6\le-2$ or $5x+2\ge22$

11. Solve for x and graph the compound solution. $3x+2>14$ or $2x-1<-7$

We solve each inequality separately.

$3x+2>14$ or $2x-1<-7$

$\quad3x>12\qquad\quad2x<-6$

$\quad\;\;x>4\qquad\quad\;\;x<-3$

The solution is $x<-3$ or $x>4$.

12. Solve for x and graph the compound solution. $4x-6\le-2$ or $5x+2\ge22$

---

13. Solve for x and graph the compound solution. $5x-1>-2$ and $3x-4<8$

We solve each inequality separately.

$5x-1>-2$ and $3x-4<8$

$\quad5x>-1\qquad\qquad3x<12$

$\quad\;\;x>-\dfrac{1}{5}\qquad\qquad x<4$

The solution is the set of numbers between $-\dfrac{1}{5}$ and 4, not including the endpoints.

$-\dfrac{1}{5}<x<4$

14. Solve for x and graph the compound solution. $2x+3\ge-5$ and $6x+4<16$

Copyright © 2013 Pearson Education, Inc.

Example	Student Practice
15. Solve. $-3x - 2 < -5$ and $4x + 6 < -12$	**16.** Solve. $2x + 3 \leq 1$ and $-6x - 10 < -34$

15. Solve. $-3x - 2 < -5$ and $4x + 6 < -12$

We solve each inequality separately.

$$-3x - 2 < -5 \quad \text{and} \quad 4x + 6 < -12$$
$$-3x < -3 \qquad\qquad 4x < -18$$
$$\frac{-3x}{-3} > \frac{-3}{-3} \qquad\qquad \frac{4x}{4} < \frac{-18}{4}$$
$$x > 1 \qquad\qquad x < -4\frac{1}{2}$$

Now, clearly it is impossible for one number to be greater than 1 and at the same time be less than $-4\frac{1}{2}$. Thus, there is no solution. We can express this by the notation $\emptyset$, which is the empty set. Or we can just state, "There is no solution."

Extra Practice

1. Graph the values of x that satisfy the conditions given. $-4 < x \leq \frac{1}{2}$

2. Graph the values of x that satisfy the conditions given. $x \leq 4$ or $x \geq \frac{13}{2}$

3. Solve for x and graph your results.
   $x - 3 \geq 2$ or $x + 1 \leq 2$

4. Express as an inequality. The length of a buttonhole b in a shirt is unacceptable if it is shorter than 0.6 inch or longer than 0.85 inch.

Concept Check
Explain why there are no values of x that satisfy these given conditions. $x + 8 < 3$ and $2x - 1 > 5$

Copyright © 2013 Pearson Education, Inc.

Chapter 2 Linear Equations and Inequalities
2.8 Absolute Value Inequalities

Vocabulary

absolute value inequality • $|ax+b|=c$ • $|ax+b|<c$ • $|ax+b|>c$

1. _____ is equivalent to $-c < ax+b < c$.

2. _____ is equivalent to $ax+b < -c$ or $ax+b > c$.

3. _____ is equivalent to $ax+b = c$ or $ax+b = -c$.

Example	Student Practice
1. Solve. $|x| \le 4.5$ The inequality $|x| \le 4.5$ means that x is less than or equal to 4.5 units from 0 on a number line. We draw a picture. Thus, the solution is $-4.5 \le x \le 4.5$.	**2.** Solve and graph. $|x| \le 3$
3. Solve and graph the solution. $|x+5| \le 10$ We want to find the values of x that make $-10 \le x+5 \le 10$ a true statement. We need to solve the compound inequality. Subtract 5 from each part. $-10-5 \le x+5-5 \le 10-5$ $\quad -15 \le x \le 5$ Thus, the solution is $-15 \le x \le 5$. We graph this solution.	**4.** Solve and graph the solution. $|x+2| \le 3$

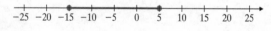

Vocabulary Answers: 1. $|ax+b|<c$ 2. $|ax+b|>c$ 3. $|ax+b|=c$

Copyright © 2013 Pearson Education, Inc.

Example	Student Practice
5. Solve and graph the solution. $\left\|x-\dfrac{2}{3}\right\| \le \dfrac{5}{2}$	**6.** Solve and graph the solution. $\left\|x-\dfrac{1}{2}\right\| < \dfrac{3}{5}$

$$-\dfrac{5}{2} \le x - \dfrac{2}{3} \le \dfrac{5}{2}$$

$$6\left(-\dfrac{5}{2}\right) \le 6(x) - 6\left(\dfrac{2}{3}\right) \le 6\left(\dfrac{5}{2}\right)$$

$$-15 \le 6x - 4 \le 15$$

$$-15 + 4 \le 6x - 4 + 4 \le 15 + 4$$

$$-11 \le 6x \le 19$$

$$-\dfrac{11}{6} \le \dfrac{6x}{6} \le \dfrac{19}{6}$$

$$-1\dfrac{5}{6} \le x \le 3\dfrac{1}{6}$$

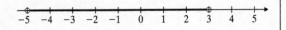

7. Solve and graph the solution.
$$|2(x-1)+4| < 8$$

First we simplify the expression within the absolute value.

$$|2x - 2 + 4| < 8$$

$$|2x + 2| < 8$$

$$-8 < 2x + 2 < 8$$

$$-8 - 2 < 2x + 2 - x < 8 - 2$$

$$-10 < 2x < 6$$

$$\dfrac{-10}{2} < \dfrac{2x}{2} < \dfrac{6}{2}$$

$$-5 < x < 3$$

8. Solve and graph the solution.
$$|3 - 2(5-x)| \le 13$$

Copyright © 2013 Pearson Education, Inc.

Example	Student Practice
9. Solve and graph the solution. $\lvert x \rvert \geq 5\frac{1}{4}$ The inequality means that x is more than $5\frac{1}{4}$ units from 0 on a number line. 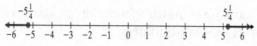 The solution is $x \leq -5\frac{1}{4}$ or $x \geq 5\frac{1}{4}$.	**10.** Solve and graph the solution. $\lvert x \rvert \geq 4$
11. Solve and graph the solution. $\lvert x - 4 \rvert > 5$ Find the values of x that make $x - 4 < -5$ or $x - 4 > 5$ a true statement. Solve each inequality separately. $x - 4 < -5$ or $x - 4 > 5$ $\quad x < -1 \qquad\qquad x > 9$ We graph the solution on a number line. 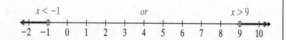	**12.** Solve and graph the solution. $\lvert x - 2 \rvert \geq 3$
13. Solve and graph the solution. $\lvert -3x + 6 \rvert > 18$ Remember to reverse the inequality sign when dividing by a negative number. $-3x + 6 > 18$ or $-3x + 6 < -18$ $\quad -3x > 12 \qquad\qquad -3x < -24$ $\quad \dfrac{-3x}{-3} < \dfrac{12}{-3} \qquad \dfrac{-3x}{-3} > \dfrac{-24}{-3}$ $\quad x < -4 \qquad\qquad x > 8$ 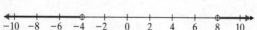	**14.** Solve and graph the solution. $\lvert -2x + 5 \rvert \geq 3$

Copyright © 2013 Pearson Education, Inc.

Example	Student Practice
15. When a new car transmission is built, the diameter d of the transmission must not differ from the specified standard s by more than 0.37 millimeter. The engineers express this requirement as $\|d - s\| \leq 0.37$. If the standard s is 216.82 millimeters for a particular car, find the limits of d.	**16.** The diameter d of a transmission must not differ from the specified standard s by more than 0.37 millimeter. This is written as $\|d - s\| \leq 0.37$. Solve to find the allowed limits of d for an SUV transmission, for which the standard s is 273.98 millimeters.

$$|d - s| \leq 0.37$$
$$|d - 216.82| \leq 0.37$$
$$-0.37 \leq d - 216.82 \leq 0.37$$
$$216.45 \leq d \leq 217.19$$

Thus the diameter of the transmission must be at least 216.45 millimeters, but not greater than 217.19 millimeters.

Extra Practice

1. Solve and graph the solution. $\|x - 3\| \leq 3$

2. Solve for x. $\|x + 3\| - 4 \leq -2$

3. Solve for x. $\left| \dfrac{3}{4}x - \dfrac{1}{4} \right| \geq 3$

4. In a certain company, the measured thickness t of a computer chip must not differ from the standard s by more than 0.05 millimeter. The engineers express this requirement as $\|t - s\| \leq 0.05$. Find the limits of t if the standard s is 1.33 millimeters.

Concept Check

Explain what happens when you try to solve for x. $\|7x + 3\| < -4$

Copyright © 2013 Pearson Education, Inc.

MATH COACH

Mastering the skills you need to do well on the test.

Watch the **MATH COACH** videos in MyMathLab®or on You Tube while you work the problems below. These helpful hints will help you avoid making common errors on test problems.

Solving Literal Equations for a Specified Variable—

Problem 5   Solve for n. $L = a + d(n-1)$

> **Helpful Hint:** Use the distributive property to remove all parentheses before continuing with the other steps of the problem.

Did you obtain the equation $L = a + dn - d$ after removing the parentheses? Yes _____ No _____

If you answered No, stop and carefully multiply each term inside the parentheses by the variable d.

You want to get the dn term on one side of the equation and all other terms on the other side.

If you added $-a + d$ to both sides, did you get $L - a + d = dn$? Yes _____ No _____

If you answered No, go back and carefully perform the step of adding $-a + d$ to each side of the equation after the parentheses have been removed.

Remember that in the final step, you must divide both sides of the equation by the variable d.

If you answered Problem 5 incorrectly, go back and rework the problem using these suggestions.

Solving Absolute Value Equations—Problem 10   Solve for x. $\left|\frac{1}{2}x+3\right| - 2 = 4$

> **Helpful Hint:** Remember that you must add a number to each side of the equation to isolate the absolute value expression.

Did you add 2 to each side of the equation to obtain $\left|\frac{1}{2}x+3\right| = 6$? Yes _____ No _____

If you answered No, stop and complete that step before you do any other operations.

In your next step, did you write the two equations $\frac{1}{2}x+3 = 6$ and $\frac{1}{2}x+3 = -6$ and solve for x in each equation? Yes _____ No _____

If you answered No, carefully perform those steps. Your answer should provide two potential solutions.

Now go back and rework the problem using these suggestions.

Copyright © 2013 Pearson Education, Inc.

Solving Linear Inequalities in One Variable—Problem 16

Solve and graph. $-\dfrac{1}{2}+\dfrac{1}{3}(2-3x) \geq \dfrac{1}{2}x+\dfrac{5}{3}$

Helpful Hint: Remove parentheses in linear inequalities as the first step. If there are fractions, multiply both sides of the inequality by the LCD to clear the fractions.

Did you first remove the parentheses to obtain the inequality

$-\dfrac{1}{2}+\dfrac{2}{3}-x \geq \dfrac{1}{2}x+\dfrac{5}{3}$?

Yes _____   No _____

If you answered No, go back and carefully multiply the terms inside the parentheses by $\dfrac{1}{3}$.

Did you multiply each term of the inequality by the LCD, 6, to obtain the inequality $-3+4-6x \geq 3x+10$?

Yes _____   No _____

If you answered No, consider why the LCD is 6 and carefully multiply all the terms of the equation by 6.

The next goal is to combine like terms on each side of the equation and get all the x terms on one side and all the numerical terms on the other side. If you divide by a negative number, you must reverse the inequality.

If you answered Problem 16 incorrectly, go back and rework the problem using these suggestions.

Solving Absolute Value Inequalities—Problem 20

Solve. $|3x+1| > 7$

Helpful Hint: If c is a positive real number and $|ax+b| > c$, then $ax+b > c$ or $ax+b < -c$. Use this first to create the two separate inequalities.

Did you apply the rule in the Helpful Hint to obtain the inequalities $3x+1 > 7$ or $3x+1 < -7$?

Yes _____   No _____

If you answered No, stop and perform those steps.

Did you add the correct number to each inequality to obtain the inequalities $3x > 6$ or $3x < -8$?

Yes _____   No _____

If you answered No, consider how to solve each inequality and perform each step carefully again.

In your final step, you will need to divide each inequality by 3 to solve for x.

Now go back and rework the problem using these suggestions.

Copyright © 2013 Pearson Education, Inc.

Name: _____   Date: _____

Instructor: _____   Section: _____

Chapter 3 Equations and Inequalities in Two Variables and Functions
3.1 Graphing Linear Equations with Two Unknowns

Vocabulary
rectangular coordinate system • x-axis • y-axis • origin • ordered pair
linear equation in two variables • standard form • solution • x-intercept
y-intercept • vertical line • horizontal line

1. The graph of the equation $x = a$, where a is any real number, is a(n) _____ through the point $(a, 0)$.

2. $Ax + By = C$ is called the _____ of a linear equation in two variables.

3. The _____ is formed by intersecting horizontal and vertical number lines.

4. An ordered pair that satisfies the equation is a(n) _____ of an equation in two variables.

Example	Student Practice
1. Graph the equation. $y = -3x + 2$	**2.** Graph the equation. $y = -2x - 1$

Make a table of x and y values.

x	y
-1	5
1	-1
2	-4

Graph the three ordered pairs and connect them with a straight line.

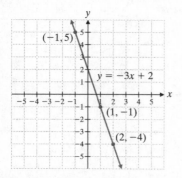

Vocabulary Answers: 1. vertical line 2. standard form 3. rectangular coordinate system 4. solution

Copyright © 2013 Pearson Education, Inc.

Example	Student Practice

3. Find the x-intercept, the y-intercept, and one additional ordered pair that satisfies the equation. Then graph the equation $4x - 3y = -12$.

4. Find the x-intercept, the y-intercept, and one additional ordered pair that satisfies the equation. Then graph the equation $3x + 2y = -6$.

Find the x-intercept by using $y = 0$ and find the y-intercept by using $x = 0$.

$$4x - 3(0) = -12 \qquad 4(0) - 3y = -12$$
$$4x = -12 \qquad\qquad -3y = -12$$
$$x = -3 \qquad\qquad\quad y = 4$$

The x-intercept is $(-3, 0)$ and the y-intercept is $(0, 4)$.

Let's pick $y = 2$ to find our third point.

$$4x - 3(2) = -12$$
$$4x - 6 = -12$$
$$x = -\frac{3}{2}$$

Hence, the third point is $\left(-\frac{3}{2}, 2\right)$.

Graph the three ordered pairs and connect them with a straight line.

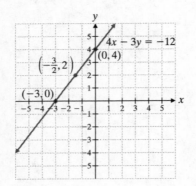

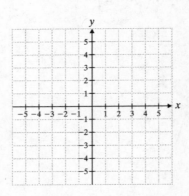

Copyright © 2013 Pearson Education, Inc.

Example	Student Practice
5. Simplify and graph the equation. $2y - 4 = 0$ The equation can be simplified to $y = 2$. For any value of x, y is 2. The graph of $y = 2$ is a horizontal line 2 units above the x-axis. 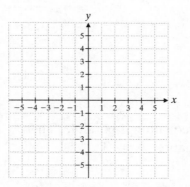	**6.** Simplify and graph the equation. $-9 = 3y$

7. A company's finance officer has determined that the monthly cost in dollars for leasing a photocopier is $C = 100 + 0.002n$, where n is the number of copies produced in a month in excess of a specified number. Graph the equation using $n = 0$, $n = 30,000$ and $n = 60,000$. Let the n-axis be the horizontal axis.

Make a table of three C and n values. Graph the three ordered pairs and connect them with a straight line.

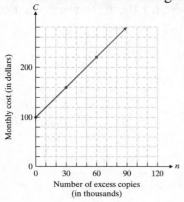

8. The cost in dollars of printing a school newspaper is $C = 75 + 0.008n$, where n is the number of copies printed in a month in excess of a specified number. Graph the equation using $n = 0$, $n = 100$ and $n = 1,000$. Let the n-axis be the horizontal axis.

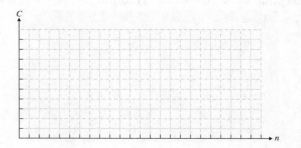

Copyright © 2013 Pearson Education, Inc.

Extra Practice

1. Graph the equation. $y = \dfrac{1}{2}x - 2$

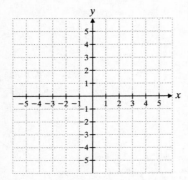

2. Simplify the equation if possible. Find the x-intercept, the y-intercept, and one additional ordered pair that is a solution to the equation. Then graph the equation. $6x + 4y + 3 = 3$

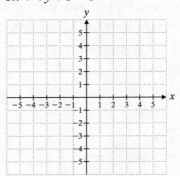

3. Simplify the equation if possible. Then graph the equation. $x = -3$

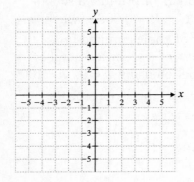

4. Simplify the equation if possible. Find the x-intercept, the y-intercept, and one additional ordered pairs that is a solution to the equation. Then graph the equation. $x - 4y = 4$

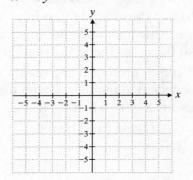

Concept Check

Explain how you would find the x-intercept and the y-intercept for the following equation. $7x - 3y = -14$

Copyright © 2013 Pearson Education, Inc.

Chapter 3 Equations and Inequalities in Two Variables and Functions
3.2 Slope of a Line

Vocabulary
rise   •   run   •   slope of a straight line   •   positive slope
negative slope   •   parallel lines   •   perpendicular lines

1. If two different lines have slopes that are equal they are said to be _____ .

2. A line sloping upward to the right has a(n) _____ .

3. In a coordinate plane, the _____ measures the ratio of the rise to the run.

4. Two lines with slopes m_1 and m_2 are _____ if $m_1 = -\dfrac{1}{m_2}$ (m_1, $m_2 \neq 0$).

Example	Student Practice
1. Find the slope of the line passing through $(-2,-3)$ and $(1,-4)$.	**2.** Find the slope of the line passing through $(9,-7)$ and $(5,3)$.

Identify the x- and y-coordinates.

$$(-2,-3) \text{ and } (1,-4)$$
$$(x_1, y_1) \qquad (x_2, y_2)$$

Use the formula.

$$\text{slope} = m = \frac{y_2 - y_1}{x_2 - x_1}$$

$$= \frac{-4 - (-3)}{1 - (-2)}$$

$$= -\frac{1}{3}$$

The slope of the line passing through $(-2,-3)$ and $(1,-4)$ is $-\dfrac{1}{3}$.

Vocabulary Answers: 1. parallel lines 2. positive slope 3. slope 4. perpendicular lines

Copyright © 2013 Pearson Education, Inc.

Example	Student Practice
3. Find the slope, if possible, of the line passing through each pair of points.	**4.** Find the slope, if possible, of the line passing through each pair of points.

3. (a) $(1.6, 2.3)$ and $(-6.4, 1.8)$

Use the slope formula and solve.

$$m = \frac{1.8 - 2.3}{-6.4 - 1.6} = \frac{-0.5}{-8.0} = 0.0625$$

(b) $\left(\dfrac{5}{3}, -\dfrac{1}{2}\right)$ and $\left(\dfrac{2}{3}, -\dfrac{1}{4}\right)$

$$m = \frac{-\dfrac{1}{4} - \left(-\dfrac{1}{2}\right)}{\dfrac{2}{3} - \dfrac{5}{3}} = \frac{-\dfrac{1}{4} + \dfrac{2}{4}}{-\dfrac{3}{3}} = -\dfrac{1}{4}$$

4. (a) $(-3.5, 0.9)$ and $(-2.1, -1.9)$

(b) $\left(\dfrac{2}{5}, \dfrac{3}{2}\right)$ and $\left(\dfrac{4}{5}, -\dfrac{1}{6}\right)$

---

5. Find the pitch of a roof as shown in the sketch. Use only positive numbers in your calculation.

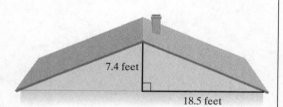

7.4 feet

18.5 feet

$$\text{slope} = \frac{\text{rise}}{\text{run}} = \frac{7.4}{18.5} = 0.4$$

This could also be expressed as the fraction $\dfrac{2}{5}$. A builder might refer to this as a pitch (slope) of $2:5$.

6. Find the pitch of a roof that has a rise of 9.6 feet and a run of 16.0 feet. Use only positive numbers in your calculation.

Copyright © 2013 Pearson Education, Inc.

Example	Student Practice
7. Find the slope of a line that is perpendicular to the line l that passes through $(4,-6)$ and $(-3,-5)$.	**8.** Find the slope of a line that is perpendicular to the line l that passes through $(-2,0)$ and $(5,4)$.

7. (continued)

First find the slope of line l.

$$m_l = \frac{-5-(-6)}{-3-4} = \frac{-5+6}{-7} = \frac{1}{-7} = -\frac{1}{7}$$

The slope of a line perpendicular to line l must have a slope of $-\dfrac{1}{m_l}$.

$$m = -\frac{1}{m_l} = -\frac{1}{\left(-\dfrac{1}{7}\right)} = -\left(-\frac{7}{1}\right) = 7$$

Example	Student Practice
9. Without plotting any points, show that the points $A(-5,-1)$, $B(-1,2)$, and $C(3,5)$ lie on the same line.	**10.** Without plotting any points, show that the points $A(0,1)$, $B(1,-1)$, and $C(3,-5)$ lie on the same line.

9. (continued)

Find the slope of the line segment from A to B and the slope of the line segment from B to C.

$$m_{AB} = \frac{2-(-1)}{-1-(-5)} = \frac{2+1}{-1+5} = \frac{3}{4}$$

$$m_{BC} = \frac{5-2}{3-(-1)} = \frac{3}{3+1} = \frac{3}{4}$$

Since the slopes are equal, we must have one line or two parallel lines. But, the line segments have a point (B) in common, so all three points line on the same line.

Copyright © 2013 Pearson Education, Inc.

Example	Student Practice
11. A gulfstream jet takes off from Orange County Airport in California. Look at the graph below. At 1 mile from the takeoff point, the jet is 1000 feet above the ground and begins a specified rate of climb. At 2 miles from the takeoff point, it is 1750 feet above the ground. At 3.5 miles from the takeoff point, it is 2865 feet above the ground. Is the jet traveling in a straight line from the 1-mile point to the 3.5-mile point?	**12.** If the jet in example **11** is 2875 feet above the ground at 3.5 miles from the takeoff point, with all other distances remaining the same, is the jet traveling in a straight line from the 1-mile point to the 3.5-mile point now?

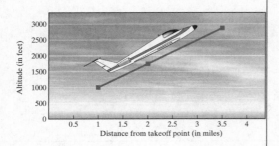

No, the slopes from the 1-mile point to the 2-mile point and from the 2-mile point to the 3.5-mile point differ.

Extra Practice

1. Find the slope, if possible, of the line passing through $(3,4)$ and $(5,-12)$.

2. Find the slope of a line parallel to the line that passes through $(-3.6, -7.2)$ and $(-7.6, -7.7)$.

3. Find the slope of a line perpendicular to the line that passes through $(0,4)$ and $\left(\frac{2}{3}, 6\right)$.

4. A river has a slope of -0.09. How many meters does it fall vertically over a horizontal distance of 800 meters?

Concept Check
How would you find the grade (slope) of a driveway that rises 8.5 feet vertically over a horizontal distance of 120 feet?

Copyright © 2013 Pearson Education, Inc.

Name: _____   Date: _____

Instructor: _____   Section: _____

Chapter 3 Equations and Inequalities in Two Variables and Functions
3.3 Graphs and the Equations of a Line

Vocabulary
slope-intercept form • y-intercept • point-slope form • parallel lines • perpendicular lines

1. The _____ is useful if only the slope of a line and a point on the line are known.

2. The _____ allows us to quickly identify the slope of a line and its y-intercept.

Example	**Student Practice**
1. Find the slope and y-intercept of the line whose graph is shown. Then use these to write an equation of the line.	**2.** Find the slope and y-intercept of the line whose graph is shown. Then use these to write an equation of the line.

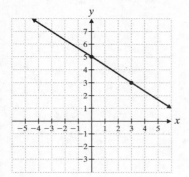 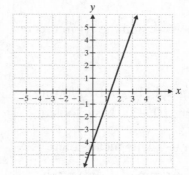

The y-intercept is at $(0,5)$ and another point on the line is $(3,3)$. Use these points to calculate the slope.

$$m = \frac{y_2 - y_1}{x_2 - x_1} = \frac{3-5}{3-0} = -\frac{2}{3}$$

Use the slope and y-intercept to write the equation in slope-intercept form.

$$y = mx + b$$

$$y = -\frac{2}{3}x + 5$$

Vocabulary Answers: 1. point-slope form 2. slope-intercept form

Copyright © 2013 Pearson Education, Inc.

Example	Student Practice
3. Find the slope and the y-intercept. Then, sketch the graph of the equation $28x - 7y = 21$.	**4.** Find the slope and the y-intercept. Then, sketch the graph of the equation $-8x + 16y = -32$.

Example

3. Find the slope and the y-intercept. Then, sketch the graph of the equation $28x - 7y = 21$.

First we will change the standard form of the equation into slope-intercept form. This is a very important procedure. Be sure that you understand each step.

$$28x - 7y = 21$$
$$-7y = -28x + 21$$
$$\frac{-7y}{-7} = \frac{-28x}{-7} + \frac{21}{-7}$$
$$y = 4x + (-3)$$

Thus, the slope is 4, and the y-intercept is $(0, -3)$.

To sketch the graph, begin by plotting the y-intercept, $(0, -3)$. Now look at the slope, 4, or $\dfrac{4}{1}$. This means there is a rise of 4 for every run of 1. From the point $(0, -3)$ go up 4 units and to the right 1 unit to locate a second point on the line. Draw the straight line that contains these two points, and you have the graph of the equation $28x - 7y = 21$.

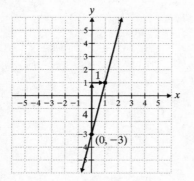

Student Practice

4. Find the slope and the y-intercept. Then, sketch the graph of the equation $-8x + 16y = -32$.

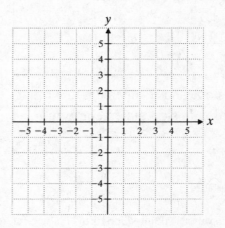

Copyright © 2013 Pearson Education, Inc.

Example	Student Practice
5. Find an equation of the line that has slope $-\dfrac{3}{4}$ and passes through the point $(-6,1)$. Express your answer in standard form.	**6.** Find an equation of the line that has slope 4 and passes through the point $(2,-4)$. Express your answer in standard form.

Since we don't know the y-intercept it is best to use the point-slope form. Substitute the given values and simplify.

$$y - y_1 = m\left(x - x_1\right)$$
$$y - 1 = -\frac{3}{4}\left[x - (-6)\right]$$
$$y - 1 = -\frac{3}{4}x - \frac{9}{2}$$

Clear the fractions and rearrange the terms to be in standard form.

$$4y - 4(1) = 4\left(-\frac{3}{4}x\right) - 4\left(\frac{9}{2}\right)$$
$$4y - 4 = -3x - 18$$
$$3x + 4y = -14$$

Example	Student Practice
7. Find an equation of the line that passes through $(3,-2)$ and $(5,1)$. Express your answer in slope-intercept form.	**8.** Find an equation of the line that passes through $(-1,-5)$ and $(4,3)$. Express your answer in slope-intercept form.

Find that the slope is $\dfrac{3}{2}$. Substitute this slope and one point into the point-slope equation and simplify.

$$y - y_1 = m\left(x - x_1\right)$$
$$y - 1 = \frac{3}{2}\left(x - 5\right)$$
$$y = \frac{3}{2}x - \frac{13}{2}$$

Copyright © 2013 Pearson Education, Inc.

Example	Student Practice
9. Find an equation of the line passing through the point $(-2, -4)$ and parallel to the line $2x + 5y = 8$. Express the answer in standard form.	**10.** Find an equation of the line passing through the point $(3, 0)$ and parallel to the line $-3x + 6y = 5$. Express the answer in standard form.

Find that the slope of the line $2x + 5y = 8$ is $-\dfrac{2}{5}$. Since parallel lines have the same slope, this is the slope of the unknown line. Substitute this and the known point into the point-slope form of the equation of a line.

$$y - y_1 = m(x - x_1)$$
$$y - (-4) = -\frac{2}{5}\left[x - (-2)\right]$$

Simplify to get $2x + 5y = -24$.

Extra Practice

1. Find the slope and y-intercept of the line whose graph is shown. Then use these to write the equation of the line.

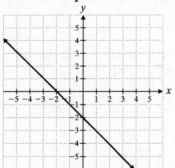

2. Write the equation in slope-intercept form. Then identify the slope and the y-intercept of the line. $3x - 5y = 15$

3. Find the equation of the line that passes through $(4, 5)$ and has slope $m = -4$. Express your answer in slope-intercept form.

4. Find the equation of the line that is perpendicular to $x + 5y = -10$ and passes through $(3, -2)$. Express your answer in standard form.

Concept Check

How would you find an equation of the horizontal line that passes through the point $(-4, -8)$?

Copyright © 2013 Pearson Education, Inc.

Name: _____   Date: _____

Instructor: _____   Section: _____

Chapter 3 Equations and Inequalities in Two Variables and Functions
3.4 Linear Inequalities in Two Variables

Vocabulary

linear equation in two variables • linear inequality in two variables • boundary line
test point • half-plane

1. A _____ is used to determine which region of a graph to shade.

2. In place of the = sign, a _____ has one of the following four inequality
   symbols: $<, <, \leq, \geq$.

Example	Student Practice
1. Graph $y < 2x + 3$.	**2.** Graph $y > -x + 2$.

Example

1. Graph $y < 2x + 3$.

The boundary line is $y = 2x + 3$.

We graph $y = 2x + 3$ using a dashed
line because the inequality contains $<$.
Since the line does not pass through
$(0,0)$, we can use it as a test point.
Substitute this point into the equation.

$$0 < 2(0) + 3$$

$$0 < 3$$

Since this inequality is true, we shade
the region on the same side of the line as
the test point. Do not include the dashed
line.

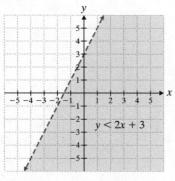

$y < 2x + 3$

Student Practice

2. Graph $y > -x + 2$.

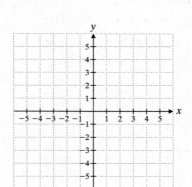

Vocabulary Answers: 1. test point 2. linear inequality in two variables

Copyright © 2013 Pearson Education, Inc.

Example	Student Practice
3. Graph $3x + 2y \geq 4$.	**4.** Graph $4x - y \leq 2$.

Graph the boundary line $3x + 2y = 4$ with a solid line. Use $(0,0)$ as the test point. Substitute $(0,0)$ into $3x + 2y \geq 4$. This inequality is false. Shade the region on the side of the line opposite $(0,0)$. The solution is the shaded region including the solid boundary line.

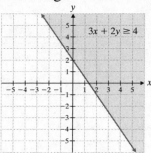

5. Graph $4x - y < 0$.

Graph the boundary line $4x - y = 0$ with a dashed line. We cannot use $(0,0)$ as a test point as it lies on the line. Choose a point not on the line and not close to it. Use $(-1,5)$. Substituting $(-1,5)$ into $4x - y < 0$ gives a true inequality. Shade the region on the side of the line that contains $(-1,5)$. The solution is the shaded region but not including the dashed boundary line.

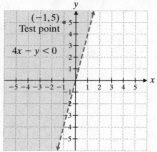

6. Graph $2x + y \geq 0$.

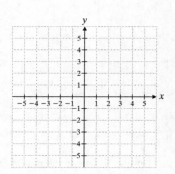

74

Copyright © 2013 Pearson Education, Inc.

Example	Student Practice
7. Simplify and graph $3y \geq -12$.	**8.** Simplify and graph $3x < 9$.

First simplify the original inequality.

$$3y \geq -12$$
$$\frac{3y}{3} \geq \frac{-12}{3}$$
$$y \geq -4$$

This inequality is equivalent to $0x + y \geq -4$. We find that if $y \geq -4$, any value of x will make the inequality true. Thus, x can be any value at all and still be included in our shaded region. Therefore, we draw a solid horizontal line at $y = 4$.

Use $(0,0)$ as a test point. Substitute $(0,0)$ into $0x + y \geq -4$ and simplify the inequality.

$$0(0) + (0) \geq -4$$
$$0 \geq -4$$

Since this inequality is true, we shade the region on the same side of the line as the test point. The solution is the shaded region including the solid boundary line.

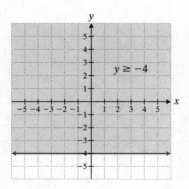

Copyright © 2013 Pearson Education, Inc.

Extra Practice

1. Graph. $y \geq \dfrac{2}{3}x + 3$

2. Graph. $-3x - y > 3$

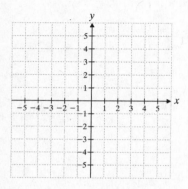

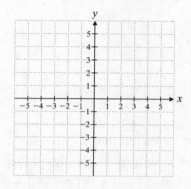

3. Graph. $5x - 3y \leq 0$

4. Graph. $2x + 2 \leq 8$

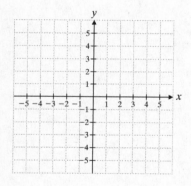

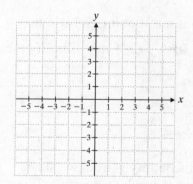

Concept Check

Explain how you would graph the following region. $4y - 8 < 0$

Copyright © 2013 Pearson Education, Inc.

Chapter 3 Equations and Inequalities in Two Variables and Functions
3.5 Concept of a Function

Vocabulary
independent variable • dependent variable • relation • domain • range
function • vertical line test • function notation

1. Any set of ordered pairs is called a(n) _____.

2. The _____ is the set of all the first items of each ordered pair in a relation.

Example	Student Practice
1. Look at the data for the men's 100-meter race. Is there a relation between any two sets of data in this table? If so, describe the relation as a table of values, a set of ordered pairs, and a graph.	**2.** Look at the data for the men's 100-meter race. Is there a relation between any two sets of data in this table? If so, describe the relation as a table of values, a set of ordered pairs, and a graph.

Olympic Games:
100-Meter Race for Men

Year	Winning runner, Country	Time
1900	Francis W. Jarvis, USA	11.0 s
1912	Ralph Craig, USA	10.8 s
1924	Harold Abrahams, Great Britain	10.6 s
1936	Jesse Owens, USA	10.3 s
1948	Harrison Dillard, USA	10.3 s
1960	Armin Harg, Germany	10.2 s
1972	Valery Borzov, USSR	10.14 s
1984	Carl Lewis, USA	9.99 s
1996	Donovan Bailey, Canada	9.84 s
2008	Usain Bolt, Jamaica	9.69 s

Source: The World Almanac

100-Meter Race for Men

Year	Course	Time
2000	Greenfield	9.52
2001	Fairview	8.08
2002	Meadows	7.75
2003	Fairview	7.05
2004	Fairview	6.42

A useful relation might be the correspondence between the year the event occurred and the winning time in the race. Choose the year as the independent variable and list it first in the table, as is customary.

Year	Time in Seconds	Year	Time in Seconds
1900	11.0	1960	10.2
1912	10.8	1972	10.14
1924	10.6	1984	9.99
1936	10.3	1996	9.84
1948	10.3	2008	9.69

Example continues on next page.

Vocabulary Answers: 1. relation 2. domain

Copyright © 2013 Pearson Education, Inc.

Example	Student Practice
Now, write the dates as ordered pairs. $\{(1900,11.0),(1912,10.8),(1924,10.6),$ $(1936,10.3),(1948,10.3),(1960,10.2),$ $(1972,10.14),(1984,9.99),(1996,9.84),$ $(2008,9.69)\}$ Finally, represent the data as a graph. 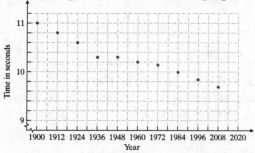	

3. Give the domain and range of the relation. Indicate whether the relation is a function.

$g = \{(2,8),(2,3),(3,7),(5,12)\}$

The domain consists of all the possible values of the independent variable or input. In a set of ordered pairs, this is the first item in each ordered pair.

Domain $= \{2,3,5\}$

The range consists of all the possible values of the dependent variable or output. In a set of ordered pairs, this is the second item in each ordered pair.

Range $= \{8,3,7,12\}$

The ordered pairs $(2,8)$ and $(2,3)$ have the same first coordinate. Thus, g is not a function.

4. Give the domain and range of the relation. Indicate whether the relation is a function.

$g = \{(4,1),(10,3),(8,3),(10,5)\}$

Copyright © 2013 Pearson Education, Inc.

Example	Student Practice

5. Give the domain and range of the relation. Indicate whether the relation is a function.

Women's Tibia Bone Lengths and Heights

Length of Tibia Bone (cm)	33	34	35	36
Height of the Woman (cm)	151	154	156	159

Source: National Center for Health Statistics

Domain $= \{33, 34, 35, 36\}$

Range $= \{151, 154, 156, 159\}$

No two different ordered pairs have the same first coordinate. This relation is a function.

6. Give the domain and range of the relation. Indicate whether the relation is a function.

A Student's Height and Age

Age	5	6	7	8
Height (in)	36	38	40	44

7. Determine whether each of the following is a graph of a function.

(a)

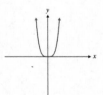

Since no vertical line will pass through more than one ordered pair on the curve, this is a function.

(b)

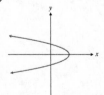

Since a vertical line could pass through two points on the curve, this is not a function.

8. Determine whether each of the following is a graph of a function.

(a)

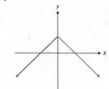

(b)

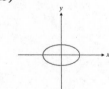

Copyright © 2013 Pearson Education, Inc.

Example	Student Practice
9. If $f(x) = 4 - 3x$, find each of the following.	**10.** If $f(x) = 10 - 2x$, find each of the following.

9. If $f(x) = 4 - 3x$, find each of the following.

(a) $f(2)$

Replace x with 2.

$$f(2) = 4 - 3(2)$$
$$= 4 - 6$$
$$f(2) = -2$$

(b) $f(-1)$

$$f(-1) = 4 - 3(-1)$$
$$= 4 + 3$$
$$f(-1) = 7$$

10. If $f(x) = 10 - 2x$, find each of the following.

(a) $f(6)$

(b) $f(-2)$

Extra Practice

1. What are the domain and range of the relation? Is the relation a function?

USA	5	6	7	8
Japan	23	24	25	26

2. Determine whether the graph represents a function.

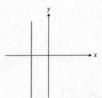

3. Given the function defined by $f(x) = 3x - 7$, find $f(0)$.

4. Given the function defined by $g(x) = 2x^2 - 3x + 5$, find $g(-3)$.

Concept Check

Explain how you can determine from a graph whether or not the graph represents a function.

Copyright © 2013 Pearson Education, Inc.

Chapter 3 Equations and Inequalities in Two Variables and Functions
3.6 Graphing Functions from Equations and Tables of Data

Vocabulary

linear function • absolute value function • symmetric about the y-axis

rational function • continuous • extrapolation • rate of change

1. A graph is said to be _____ when the graph on one side of the y-axis is a mirror image of the graph on the other side of the y-axis.

2. The graph of a(n) _____ is a straight line.

Example	Student Practice
1. Graph. $f(x) = |x|$	**2.** Graph. $g(x) = |x| + 1$

Choose both negative and positive values for x.

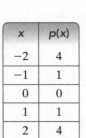
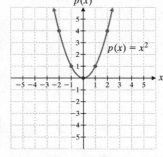
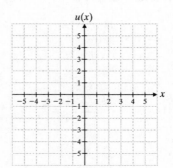

x	f(x)
−2	2
−1	1
0	0
1	1
2	2

3. Graph. $p(x) = x^2$ **4.** Graph. $u(x) = x^2 - 1$

Use both negative and positive values of x. Since this is not a linear function, use a curved line to connect the points.

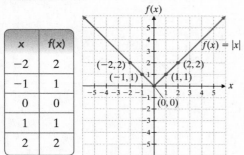
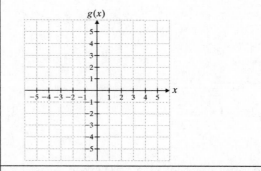

x	p(x)
−2	4
−1	1
0	0
1	1
2	4

Vocabulary Answers: 1. symmetric about the y-axis 2. linear function

Copyright © 2013 Pearson Education, Inc.

Example	Student Practice
5. Graph. $g(x) = x^3$	**6.** Graph. $h(x) = x^3 - 1$

5. Graph. $g(x) = x^3$

Use five values for x, find their function values, and plot the five points to assist in sketching the graph.

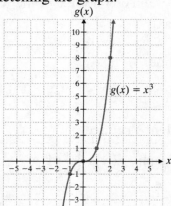

x	g(x)
−2	−8
−1	−1
0	0
1	1
2	8

6. Graph. $h(x) = x^3 - 1$

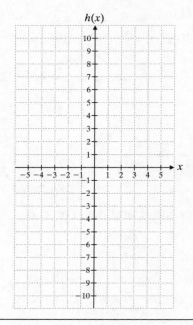

7. Graph. $p(x) = \dfrac{4}{x}$

We cannot choose x to be 0 since $\dfrac{4}{0}$ is not defined. The domain of this function is all real numbers except 0. Use five values for x that are greater than 0 and five values that are less than 0.

x	p(x)
8	$\frac{1}{2}$
4	1
2	2
1	4
$\frac{1}{2}$	8
$-\frac{1}{2}$	−8
−1	−4
−2	−2
−4	−1
−8	$-\frac{1}{2}$

8. Graph. $r(x) = \dfrac{2}{x}$

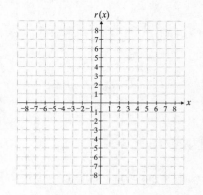

Copyright © 2013 Pearson Education, Inc.

Example	Student Practice

9. The marketing manager of a shoe company compiled the data in the table.

x	p(x)
Pairs of Shoes Sold in a Month (In Thousands of Pairs)	**Monthly Profit from the Sales of Shoes (In Thousands of Dollars)**
3	5
5	9
7	13

(a) Plot the data values and connect the points to see the graph of the underlying function.

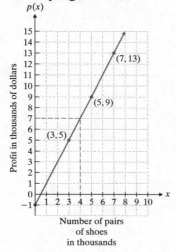

(b) Determine the profit from selling 4000 pairs of shoes in one month.

We find that the value $x = 4$ on the graph corresponds to the value $y = 7$. If we sold 4000 pairs of shoes, we would have a profit of $7000.

(c) What value do you obtain on the graph for $x = 0$? What does this mean?

We find that when $x = 0$, $y = -1$. This would predict a loss of $1000 in a month of zero sales of shoes.

10. A manager of a restaurant compiled the data in the table.

x	p(x)
Average Number of Customers Served Per Hour (In Tens of Customers)	**Weekly Profit (In Hundreds of Dollars)**
3	2
6	6
9	10

(a) Plot the data values and connect the points to see the graph of the underlying function.

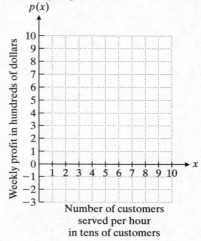

(b) Determine the profit if the restaurant serves 50 customers per hour.

(c) What kind of profit would you expect to make from serving 0 customers per hour?

Copyright © 2013 Pearson Education, Inc.

Example	Student Practice

11. Graph the function suggested by the data given in the table. The domain is $x \geq 2$, and the range is $f(x) \geq 0$.

Find an approximate value for $f(x)$ when $x = 8$.

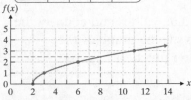

x	2	3	6	11
f(x)	0	1	2	3

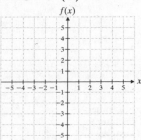

When $x = 8$, $f(x)$ is about 2.5.

12. Graph the function suggested by the data given in the table. The domain is $x \geq 1$, and the range is $f(x) \geq 2$.

Find an approximate value for $f(x)$ when $x = 3$.

x	1	2	5	10
f(x)	2	3	4	5

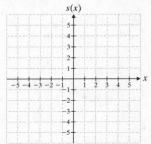

Extra Practice

1. Graph $f(x) = 2x - 2$.

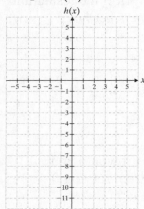

2. Graph $s(x) = |x + 3|$.

3. Graph $h(x) = -x^3 - 3$.

4. Graph the function based on the table of values. Assume that both the domain and range are all real numbers.

x	f(x)
−2	−4
0	−1
2	2

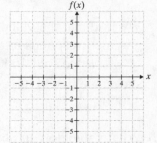

Concept Check

Explain how you would find $f\left(\dfrac{1}{4}\right)$ for the following function. $f(x) = \dfrac{5}{8x - 5}$

Copyright © 2013 Pearson Education, Inc.

MATH COACH

Mastering the skills you need to do well on the test.

Watch the **MATH COACH** videos in MyMathLab® or on You Tube™ while you work the problems below. These helpful hints will help you avoid making common errors on test problems.

Graphing a Linear Equation—Problem 4
Graph the line. Plot at least three points. $2x + 3y = -10$

> **Helpful Hint:** If you carefully plot three points that form ordered-pair solutions to the given equation and then draw a straight line through these points, you can construct the graph of the line. If you cannot connect the three points with a straight line, then you know you have made a mistake.

Did you let $y = 0$ and substitute for y in the original equation to obtain -5 as the value for x? Did you plot $(-5, 0)$ on your graph paper? Yes _____ No _____
If you answered No to either of these questions, substitute $y = 0$ into the original equation and solve for x.
Plot your results for that ordered pair on your graph paper.

Did you let $x = -2$ and substitute for x in the original equation to obtain -2 as the value for y?
Yes _____ No _____
Did you let $x = 1$ in the original equation and solve for y to get $y = -4$? Yes _____ No _____

If you answered No to either of these questions, substitute these values for x into the original equation and solve for y in each case. Then plot each of these ordered pairs.

Remember to carefully draw a straight line that passes through each of the three points.

If you answered Problem 4 incorrectly, go back and rework the problem using these suggestions.

Write the Equation of a Line Given Two Points on the Line—Problem 9
Write the standard form of the equation of the line that passes through $(5, -2)$ and $(-3, -1)$.

> **Helpful Hint:** First, we can find the slope of a line when given two points on that line. Then we substitute the slope and the coordinates of one of the given points into the point-slope equation. Remember to simplify the equation if necessary.

Did you obtain a slope of $-\dfrac{1}{8}$ when you substituted the values of x and y for each point in the slope formula?
Yes _____ No _____
If you answered No, carefully substitute the x and y values into the slope formula.

Did you correctly substitute the slope and the point $(-3, -1)$ or $(5, -2)$ into the point-slope equation to obtain either

$y + 1 = -\dfrac{1}{8}(x + 3)$ or $y + 2 = -\dfrac{1}{8}(x - 5)$?
Yes _____ No _____

If you answered No, please carefully write the point-slope equation and substitute either the point $(-3, -1)$ or $(5, -2)$ to see if you can obtain one of these two equations. In your final step, remember to simplify the equation and write it in standard form.

Now go back and rework the problem using these suggestions.

Copyright © 2013 Pearson Education, Inc.

Graphing a Linear Inequality in Two Variables—Problem 13   Graph the region. $4x - 2y < -6$

Helpful Hint: Remember to use a dashed line for the graph of the boundary line if the inequality is written with a > or a < symbol.

Did you draw a dashed line for the graph of the boundary line $4x - 2y = -6$? Did it pass through the points $(0,3)$ and $(-3,-3)$?

Yes _____ No _____

If you answered No, remember to carefully plot points along the boundary line and connect these points with a dashed line.

If you use the test point $(0,0)$, does it satisfy the inequality $4x - 2y < -6$?

Yes _____ No _____

If you answered Yes, this is not correct. When you substitute $(0,0)$ into the inequality, you obtain $0 < -6$, which is not true. Does this mean that we shade above or below the dashed line? Remember that we shade the side of the boundary line where any test points chosen make the inequality true.

If you answered Problem 13 incorrectly, go back and rework the problem using these suggestions.

Evaluating a Function Using Function Notation—Problem 18
If $p(x) = -2x^3 + 3x^2 + x - 4$, find $p(-2)$.

Helpful Hint: When you replace x with a numerical value in the function, place parentheses around the number in the substitution step to avoid making mathematical errors.

Did you obtain $p(-2) = -2(-2)^3 + 3(-2)^2 + (-2) - 4$ in your first step?

Yes _____ No _____

If you answered No, stop and perform that step carefully again. Writing out this step will help you avoid errors.

Did you obtain $p(-2) = -2(-8) + 3(4) + (-2) - 4$ in your next step?

Yes _____ No _____

If you answered No, apply the order of operations by evaluating exponents next. Note that -2 to the third power is $(-2)(-2)(-2) = -8$ and -2 squared is $(-2)(-2) = 4$.

Now go back and rework the problem using these suggestions.

Copyright © 2013 Pearson Education, Inc.

Chapter 4 Systems of Linear Equations and Inequalities
4.1 Systems of Linear Equations in Two Variables

Vocabulary

systems of two linear equations in two variables   •   solution to a system   •   inconsistent
dependent   •   substitution method   •   addition method   •   independent
consistent system   •   no solution   •   identity

1. If a system of equations has no solution, it is said to be _____.

2. In the _____, we choose one equation and solve for one variable. Then we
    substitute this expression into the other equation.

3. A system with infinitely many solutions is said to be a(n) _____.

Example	Student Practice
1. Determine whether $(3, -2)$ is a solution to the following system.	**2.** Determine whether $(3, 2)$ is a solution to the following system.

Example:

$$x + 3y = -3$$
$$4x + 3y = 6$$

Substitute $(3, -2)$ into the first equation
to see whether the ordered pair is a
solution to the first equation.

$$(3) + 3(-2) \overset{?}{=} -3$$

$$3 - 6 \overset{?}{=} -3$$

$$-3 = -3$$

Likewise, determine whether $(3, -2)$ is
a solution to the second equation.

$$4(3) + 3(-2) \overset{?}{=} 6$$

$$6 = 6$$

Since $(3, -2)$ is a solution to each
equation, it is a solution to the system.

Student Practice:

$$2x + y = 8$$
$$3x + 5y = 19$$

Vocabulary Answers: 1. inconsistent 2. substitution method 3. dependent

Copyright © 2013 Pearson Education, Inc.

Example	Student Practice

3. Solve this system of equations by graphing.

$$2x + 3y = 12$$
$$x - y = 1$$

Graph each line. Then, find the point of intersection.

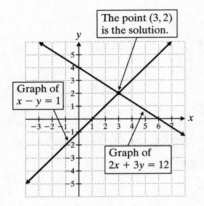

The point $(3, 2)$ is the solution.

Graph of $x - y = 1$

Graph of $2x + 3y = 12$

The solution is $(3, 2)$. The graphing method does not always lead to an accurate result because it involves a visual estimation of the point of intersection. Verify the solution by substituting $(3, 2)$ into each equation.

$$2(3) + 3(2) \overset{?}{=} 12 \qquad 3 - 2 \overset{?}{=} 1$$
$$12 = 12 \qquad\qquad 1 = 1$$

Thus, we have verified that the solution to the system is $(3, 2)$.

4. Solve this system of equations by graphing.

$$-x + y = -6$$
$$3x + 5y = 2$$

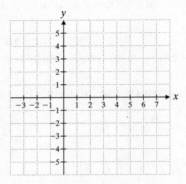

Copyright © 2013 Pearson Education, Inc.

Example	Student Practice
5. Find the solution to the following system of equations. Use the substitution method.	**6.** Find the solution to the following system of equations. Use the substitution method.

Example (left column):

$$x + 3y = -7 \quad (1)$$
$$4x + 3y = -1 \quad (2)$$

Solve for x in equation (1) because x has a coefficient of 1, $x = -7 - 3y$ (3). Now, substitute this expression for x in equation (2) and solve for y.

$$4x + 3y = -1$$
$$4(-7 - 3y) + 3y = -1$$
$$-28 - 12y + 3y = -1$$
$$-28 - 9y = -1$$
$$-9y = 27$$
$$y = -3$$

Substitute $y = -3$ into equation (1) to find x.

$$x + 3(-3) = -7$$
$$x - 9 = -7$$
$$x = 2$$

The solution is $(2, -3)$. Verify the solution in both of the original equations.

$$2 + 3(-3) \overset{?}{=} -7 \qquad 4(2) + 3(-3) \overset{?}{=} -1$$
$$2 - 9 \overset{?}{=} -7 \qquad\qquad 8 - 9 \overset{?}{=} -1$$
$$-7 = -7 \qquad\qquad\qquad -1 = -1$$

The solution checks.

Student Practice (right column):

$$x - 2y = 4 \quad (1)$$
$$2x - 5y = 5 \quad (2)$$

Copyright © 2013 Pearson Education, Inc.

Example	Student Practice
7. Solve the following system by the addition method.	**8.** Solve the following system by the addition method.

$$\frac{x}{4} + \frac{y}{6} = -\frac{2}{3} \quad (1)$$

$$\frac{x}{5} + \frac{y}{2} = \frac{1}{5} \quad (2)$$

$$\frac{x}{6} + \frac{y}{9} = \frac{11}{9} \quad (1)$$

$$\frac{x}{3} + \frac{y}{7} = \frac{43}{21} \quad (2)$$

Clear equation (1) of fractions by multiplying each term by 12. Clear equation (2) of fraction by multiplying each term by 10. Now we have an equivalent system that does not contain fractions.

$$3x + 2y = -8 \quad (3)$$
$$2x + 5y = 2 \quad (4)$$

To eliminate the variable x, we multiply equation (3) by 2 and (4) by 3. We now have the following equivalent system. Add the equations, then solve for y.

$$6x + 4y = -16$$
$$\underline{-6x - 15y = -6}$$
$$-11y = -22$$
$$y = 2$$

Substitute $y = 2$ into equation (3) and solve for x.

$$3x + 2(2) = -8$$
$$3x + 4 = -8$$
$$3x = -12$$
$$x = -4$$

The solution to the system is $(-4, 2)$.
The check is left to the student.

Copyright © 2013 Pearson Education, Inc.

Example	Student Practice
9. If possible, solve the system. $\quad 2x + 8y = 16 \quad (1)$ $\quad 4x + 16y = -8 \quad (2)$ To eliminate y, we'll multiply equation (1) by -2. $-2(2x) + (-2)(8y) = (-2)(16)$ $\qquad -4x - 16y = -32 \qquad (3)$ We now have the following equivalent system. $-4x - 16y = -32 \quad (3)$ $\quad 4x + 16y = -8 \quad (2)$ When we add equations (3) and (2), we get $0 = -40$, which is false. Thus, we conclude that this system is inconsistent, and there is no solution.	**10.** If possible, solve the system. $\quad x + 4y = 5 \quad (1)$ $\quad -3x - 12y = 15 \quad (2)$
11. If possible, solve the system. $\quad 0.5x - 0.2y = 1.3 \quad (1)$ $\quad -1.0x + 0.4y = -2.6 \quad (2)$ Clear the decimals by multiplying the equations by 10. $\quad 5x - 2y = 13 \quad (3)$ $-10x + 4y = -26 \quad (4)$ Eliminate y by multiplying each term of equation (3) by 2. $\quad 10x - 4y = 26 \quad (5)$ $\underline{-10x + 4y = -26 \quad (4)}$ $\qquad 0 = 0$ This resulting statement is an identity. The equations are dependent and there is an infinite number of solutions.	**12.** If possible, solve the system. $\quad 1.4x + 0.6y = -7.4 \quad (1)$ $\quad -0.7x - 0.3y = 3.7 \quad (2)$

Copyright © 2013 Pearson Education, Inc.

Example	Student Practice

13. Select a method and solve each system of equations.

(a)
$$x + y = 3080$$
$$2x + 3y = 8740$$

Since there are x- and y-variables that have coefficients of 1, select the substitution method. The solution to the system is $(500, 2580)$. The check is left to the student.

(b)
$$5x - 2y = 19$$
$$-3x + 7y = 35$$

Since none of the x- or y-variables have coefficients of 1 or -1, select the addition method. The solution to the system is $(7, 8)$. The check is left to the student.

14. Select a method and solve each system of equations.

(a)
$$x + y = 2235$$
$$-4x + 3y = 2155$$

(b)
$$4x + 7y = 2$$
$$-5x - 8y = -4$$

Extra Practice

If possible, solve the system. If there is not a unique solution to a system, state a reason.

1.
$$3x + y = -1$$
$$4x + 2y = 0$$

2.
$$y = 3x + 2$$
$$-15x + 5y = 0$$

3.
$$6x + 3y = 12$$
$$6x - 3y = -12$$

4.
$$10y - 2x = 8$$
$$y = \frac{1}{5}x + \frac{4}{5}$$

Concept Check

Explain what happens when you go through the steps to solve the following system of equations. Why does this happen?

$$6x - 4y = 8$$
$$-9x + 6y = -12$$

Copyright © 2013 Pearson Education, Inc.

Chapter 4 Systems of Linear Equations and Inequalities
4.2 Systems of Linear Equations in Three Variables

Vocabulary

systems of three linear equations in three variables   •   solution
ordered triple   •   substitution method   •   addition method

1. The solution to a system of three linear equations in three unknowns is called a(n) _____.

2. To solve a system of three linear equations in three variables, first use the _____ to eliminate any variable from any pair of equations.

Example	Student Practice
1. Determine whether $(2,-5,1)$ is a solution to the following system. $3x + y + 2z = 3$ $4x + 2y - z = -3$ $x + y + 5z = 2$	**2.** Determine whether $(2,8,-7)$ is a solution to the following system. $4x - y + 3z = -21$ $x - y - 2z = 8$ $14x - 3y + z = -3$

Substitute $x = 2$, $y = -5$, and $z = 1$ into the first equation to see whether the ordered triple is a solution to the first equation.

$$3(2) + (-5) + 2(1) \overset{?}{=} 3$$

$$6 - 5 + 2 \overset{?}{=} 3$$

$$3 = 3$$

Likewise, determine whether $(2,-5,1)$ is a solution to the second and third equations.

$$4(2) + 4(-5) - 1 \overset{?}{=} -3 \qquad 2 + (-5) + 5(1) \overset{?}{=} 2$$

$$8 - 10 - 1 \overset{?}{=} -3 \qquad 2 - 5 + 5 \overset{?}{=} 2$$

$$-3 = -3 \qquad 2 = 2$$

Since $(2,-5,1)$ is a solution to each equation, it is a solution to the system.

Vocabulary Answers: 1. ordered triple 2. addition method

Copyright © 2013 Pearson Education, Inc.

Example	Student Practice

3. Find the solution to (that is, solve) the following system of equations.

$$-2x+5y+z=8 \quad (1)$$
$$-x+2y+3z=13 \quad (2)$$
$$x+3y-z=5 \quad (3)$$

Add equations (1) and (3) to eliminate z.

$$-2x+5y+z=8 \quad (1)$$
$$\underline{x+3y-z=5} \quad (3)$$
$$-x+8y=13 \quad (4)$$

Now, choose a different pair of the original equations and eliminate the same variable. Multiply equation (3) by 3 and call it equation (6). Add the result to equation (2).

$$-x+2y+3z=13 \quad (2)$$
$$\underline{3x+9y-3z=15} \quad (6)$$
$$2x+11y=28 \quad (5)$$

Solve the resulting system of two linear equations in two unknowns.

$$-x+8y=13 \quad (4)$$
$$2x+11y=28 \quad (5)$$

Thus, $x=3$ and $y=2$. Substitute these values into one of the original equations to find z.

$$-2(3)+5(2)+z=8$$
$$-6+10+z=8$$
$$z=4$$

The solution to the system is $(3,2,4)$.

The check is left to the student.

4. Find the solution to (that is, solve) the following system of equations.

$$3x+2y+9z=10 \quad (1)$$
$$4x-2y-z=2 \quad (2)$$
$$4x+5y+12z=32 \quad (3)$$

Copyright © 2013 Pearson Education, Inc.

Example	Student Practice
5. Solve the system.	**6.** Solve the system.

5. Solve the system.

$$4x + 3y + 3z = 4 \quad (1)$$
$$3x \quad\quad + 2z = 2 \quad (2)$$
$$2x - 5y \quad\quad = -4 \quad (3)$$

Use equations (2) and (1) to obtain an equation that contains only x and y. Multiply equation (1) by 2 and equation (2) by -3 to obtain the following system.

$$8x + 6y + 6z = 8 \quad (4)$$
$$\underline{-9x \quad\quad - 6z = -6} \quad (5)$$
$$-x + 6y \quad\quad = 2 \quad (6)$$

Notice that equation (3) already has no z-term. Solve the system formed by equations (3) and (6).

$$2x - 5y = -4 \quad (3)$$
$$-x + 6y = 2 \quad (6)$$

Thus, $x = 3$ and $y = 2$. Substitute these values into one of the original equations containing z. Use equation (2) since it only has two variables.

$$3x + 2z = 2$$
$$3(-2) + 2z = 2$$
$$2z = 8$$
$$z = 4$$

The solution to the system is $(-2, 0, 4)$. The check is left to the student.

6. Solve the system.

$$5x + 2y + 3z = -16 \quad (1)$$
$$3x + 7y \quad\quad = 11 \quad (2)$$
$$9x \quad\quad - 2z = 1 \quad (3)$$

Copyright © 2013 Pearson Education, Inc.

Extra Practice

1. Solve the system.

$$x + 4y - z = -15$$
$$2x - y - 2z = -12$$
$$3x - y + z = 1$$

2. Solve the system.

$$0.2x + 0.4y + 0.6z = 0$$
$$0.1x + y - 0.5z = -1.6$$
$$0.2x - 0.5y + 0.1z = 0.4$$

3. Solve the system.

$$x - y = 0$$
$$y - z = 5$$
$$x + y + z = 13$$

4. Solve the system, if possible.

$$x + y - z = 4$$
$$2x - 5y + z = 1$$
$$3x + 3y - 3z = 0$$

Concept Check

Explain how you would eliminate the variable z and obtain two equations with only the variables x and y in the following system.

$$2x + 4y - 2z = -22$$
$$4x + 3y + 5z = -10$$
$$5x - 2y + 3z = 13$$

Copyright © 2013 Pearson Education, Inc.

Name: _____   Date: _____

Instructor: _____   Section: _____

Chapter 4 Systems of Linear Equations and Inequalities
4.3 Applications of Systems of Linear Equations

Vocabulary
understand the problem   • $D = RT$ •   check   •   write a system of equations

1. When solving applied problems using equations, the first step is to _____.

2. The final step in the problem solving process is to _____ the answer.

Example	Student Practice
1. For the paleontology lecture on campus, advance tickets cost $5 and tickets at the door cost $6. The ticket sales this year came to $4540. The department chairman wants to raise prices next year to $7 for advance tickets and $9 for tickets at the door. He said that if exactly the same number of people attend next year, the ticket sales at these new prices will total $6560. If he is correct, how many tickets were sold in advance this year? How many tickets were sold at the door?	**2.** For the concert on campus, advance tickets cost $15 and tickets at the door cost $20. The ticket sales this year came to $6250. The activities coordinator wants to raise prices next year to $17 for advance tickets and $25 for tickets at the door. She said that if exactly the same number of people attend next year, the ticket sales at these new prices will total $7375. If she is correct, how many tickets were sold in advance this year? How many tickets were sold at the door?

Let $x =$ the number of tickets bought in advance and $y =$ the number of tickets bought at the door. The total sales for advance tickets will be $5x$ and for door tickets, $6y$. Thus, we have $5x + 6y = 4540$. The equation for next year's sales is $7x + 9y = 6560$.

$$5x + 6y = 4540$$
$$7x + 9y = 6560$$

Solve the system to find that 500 advance tickets were sold and 340 door tickets were sold. The check is left to the student.

Vocabulary Answers: 1. understand the problem 2. check

Copyright © 2013 Pearson Education, Inc.

Example	Student Practice

3. An airplane travels between two cities that are 1500 miles apart. The trip against the wind takes 3 hours. The return trip with the wind takes $2\dfrac{1}{2}$ hours. What is the speed of the plane in still air (in other words, how fast would the plane travel if there were no wind)? What is the speed of the wind?

Let $x =$ the speed of the plane in still air and let $y =$ the speed of the wind. The wind speed opposes the plane's speed in still air, so we must subtract, $x - y$. The wind speed is added to the planes speed in still air, and we add, $x + y$. Using the formula $(\text{rate})(\text{time}) = \text{distance},$ we have the equations $(x - y)(3) = 1500$ and $(x + y)(2.5) = 1500.$ Remove parentheses to get the following system.
$$3x - 3y = 1500 \quad (1)$$
$$2.5x + 2.5y = 1500 \quad (2)$$

Solve the system using the addition method.
$$15x - 15y = 7500$$
$$\underline{15x + 15y = 9000}$$
$$30x \qquad = 16{,}500$$
$$x = 550$$

Substitute this result into equation (1) to find $y.$
$$3(550) - 3y = 1500$$
$$y = 50$$
The speed of the plane in still air is 550 miles per hour and the speed of the wind is 50 miles per hour.

4. A boat travels upstream 40 miles. The trip against the current takes 4 hours. The return trip with the current takes 2 hours. What is the speed of the boat in still water (in other words, how fast would the boat travel if there was no current)? What is the speed of the current?

Copyright © 2013 Pearson Education, Inc.

Example	Student Practice

5. A trucking firm has three sizes of trucks. The biggest truck holds 10 tons of gravel, the next size holds 6 tons, and the smallest holds 4 tons. The firm has a contract to provide 15 trucks to haul 104 tons of gravel. To reduce fuel costs the firm's manager wants to use two more of the fuel-efficient 10-ton trucks than the 6-ton trucks. How many trucks of each type should she use?

6. A trucking firm has three sizes of trucks. The biggest truck holds 12 tons of gravel, the next size holds 8 tons, and the smallest holds 6 tons. The firm has a contract to provide 16 trucks to haul 154 tons of gravel. To reduce fuel costs the firm's manager wants to use three more of the fuel-efficient 12-ton trucks than the 8-ton trucks. How many trucks of each type should he use?

Let $x =$ the number of 10-ton trucks used, $y =$ the number of 6-ton trucks used, and $z =$ the number of 4-ton trucks used. Fifteen trucks will be used. So, $x + y + z = 15$. There are 104 tons of cargo to be carried and the different trucks carry 10, 6, and 4 tons, respectively. So, $10x + 6y + 4z = 104$. Two more 10-ton trucks than 6-ton trucks are used. So, $x - y = 2$. The system is as follows.

$$x + y + z = 15 \quad (1)$$
$$10x + 6y + 4z = 104 \quad (2)$$
$$x - y = 2 \quad (3)$$

Eliminate z using equations (1) and (2). Use the result and equation (3) to find x.

$$3x + y = 22$$
$$\underline{x - y = 2}$$
$$4x = 24$$
$$x = 6$$

Use the result to find y and then z using equations (3) and (1).

$$6 - y = 2 \qquad 6 + 4 + z = 15$$
$$\text{and}$$
$$y = 4 \qquad\qquad z = 5$$

The manager needs six 10-ton trucks, four 6-ton trucks, and five 4-ton trucks.

99

Copyright © 2013 Pearson Education, Inc.

Extra Practice

1. The sum of two numbers is 102. If three times the smaller number is subtracted from twice the larger number, the result is 49. Find the two numbers.

2. The Revel family farm has 500 acres of land. It costs $60 to plant an acre of soybeans and $36 to plant an acre of corn. If the Revels want to spend a total of $22,440 on planting, how many acres of each crop should they plant?

3. Devon bought 15 items at the office supply store. She spent a total of $26.75. The binders cost $2.40, the pens cost $1.85, and the erasers cost $0.60. Devon bought 4 more pens than erasers. How many of each item did she buy?

4. A total of 250 people attended a movie. The tickets cost $11 for adults, $8 for students, and $7 for senior citizens. The ticket sales totaled $2318. The manager found that if they had raised the prices to $14 for adults, $10 for students, and $8 for senior citizens, they would have made $2892. How many tickets of each type were sold?

Concept Check

A plane flew 1200 miles with a tail wind in 2.5 hours. The return trip against the wind took 3 hours. This situation is represented by the following system. Explain how you would set up two equations using the given information if the plane flew 1500 miles instead of 1200 miles.

$$2.5x + 2.5y = 1200$$
$$3x - 3y = 1200$$

Copyright © 2013 Pearson Education, Inc.

Name: _____   Date: _____

Instructor: _____   Section: _____

Chapter 4 Systems of Linear Equations and Inequalities
4.4 Systems of Linear Inequalities

Vocabulary
system of linear inequalities in two variables   •   vertex   •   intersection

1. The solution to a system of inequalities is the _____ of the solution sets of the individual inequalities of the system.

2. We call two linear inequalities in two variables a(n) _____.

3. In the solution to a system of linear inequalities, a point where the boundary lines intersect is called a(n) _____ of the solution.

Example	**Student Practice**
1. Graph the solution to the system.	**2.** Graph the solution to the system.

Example

1. Graph the solution to the system.

$$y \le -3x + 2$$
$$-2x + y \ge -1$$

The graph of $y \le -3x + 2$ is the region on and below the line $y = -3x + 2$. The graph of $-2x + y \ge -1$ is the region on and above the line $-2x + y = -1$. The two solutions are graphed on one rectangular coordinate system below. The darker shaded region is the intersection of the two graphs. Thus, the solution to the system of two inequalities is the darker shaded region and its boundary lines.

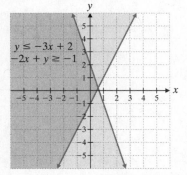

Student Practice

2. Graph the solution to the system.

$$y \le -x + 4$$
$$-5x + y \ge -1$$

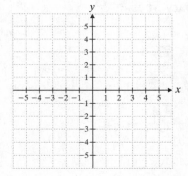

Vocabulary Answers: 1. intersection 2. system of linear inequalities in two variables 3. vertex

Copyright © 2013 Pearson Education, Inc.

Example	Student Practice
3. Graph the solution to the system.	**4.** Graph the solution to the system.

3. Graph the solution to the system.

$$y < 4$$

$$y > \frac{3}{2}x - 2$$

The graph of $y < 4$ is the region below the line $y = 4$. It does not include the line since the inequality symbol is $<$. Thus, we use a dashed line to indicate that the boundary line is not part of the answer. The graph of $y > \frac{3}{2}x - 2$ is the region above the line $y = \frac{3}{2}x - 2$. Again, we use a dashed line to indicate that the boundary line is not part of the answer. The final solution is the darker shaded region. The solution does not include the dashed boundary lines.

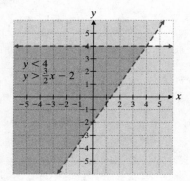

4. Graph the solution to the system.

$$y > -3$$

$$y < -\frac{2}{3}x - 3$$

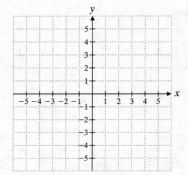

Copyright © 2013 Pearson Education, Inc.

Example	Student Practice
5. Graph the solution of the following system of inequalities. Find the coordinates of any points where boundary lines intersect.	**6.** Graph the solution of the following system of inequalities. Find the coordinates of any points where boundary lines intersect.

$$x + y \le 5$$
$$x + 2y \le 8$$
$$x \ge 0$$
$$y \ge 0$$

The graph of $x + y \le 5$ is the region on and below the line $x + y = 5$. The graph of $x + 2y \le 8$ is the region on and below the line $x + 2y = 8$. We solve the system containing the equations $x + y = 5$ and $x + 2y = 8$ to find that their point of intersection is $(2,3)$. The graph of $x \ge 0$ is the y-axis and all the region to the right of the y-axis. The graph of $y \ge 0$ is the x-axis and all the region above the x-axis. Thus, the solution to the system is the shaded region and its boundary lines. There are four points where the boundary lines intersect. These points are called the vertices of the solution. Thus, the vertices of the solution are $(0,0)$, $(0,4)$, $(2,3)$, and $(5,0)$.

$$x + y \le 5$$
$$x + 3y \le 9$$
$$x \ge 0$$
$$y \ge 0$$

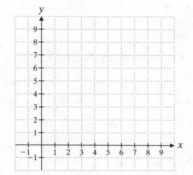

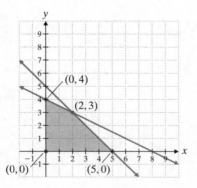

Copyright © 2013 Pearson Education, Inc.

Extra Practice

1. Graph the solution for the following system.

$$4x + 2y \geq -4$$
$$y > x - 3$$

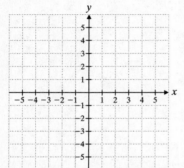

2. Graph the solution for the following system.

$$x < -3$$
$$y \geq -5$$

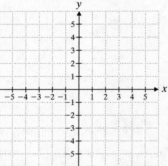

3. Graph the solution for the following system.

$$y + x \leq 4$$
$$y + x > 1$$

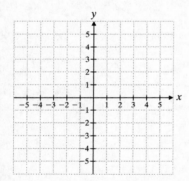

4. Graph the solution to the following system of inequalities. Find the vertices of the solution.

$$y > -3x - 5$$
$$y < 4$$
$$-2x + y > 0$$

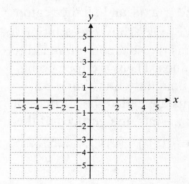

Concept Check

Explain how you would graph the region described by the following.

$$y > x + 2$$
$$x < 3$$

Copyright © 2013 Pearson Education, Inc.

MATH COACH

Mastering the skills you need to do well on the test.

Watch the **MATH COACH** videos in MyMathLab°or on You Tube™
while you work the problems below. These helpful hints will
help you avoid making common errors on test problems.

Solving a System of Linear Equations with Fractions—Problem 5

Solve the system. $\dfrac{1}{3}x + \dfrac{5}{6}y = 2$

$\qquad\qquad \dfrac{3}{5}x - y = -\dfrac{7}{5}$

> **Helpful Hint:** First clear the fractions by multiplying the first equation by
> its LCD and multiplying the second equation by its LCD. Then choose the
> appropriate method to solve this system.

Did you multiply the first equation by the LCD, 6? Did you
obtain the equation $2x + 5y = 12$? Yes _____ No _____

If you answered No, consider why 6 is the LCD and perform
the multiplication again. Remember to multiply each term of
the equation by the LCD.

Did you multiply the second equation by 5? Did you obtain
the equation $3x - 5y = -7$? Yes _____ No _____

If you answered No, go back and perform those
multiplications again. Be careful as you carry out
calculations.

Did you use the addition (elimination) method
to complete the solution?

Yes _____ No _____

If you answered No, consider why this method
might make the solution easiest to find.

If you answered Problem 5 incorrectly, go
back and rework the problem using these
suggestions.

Solving a System of Three Linear Equations in Three Variables—Problem 7   Solve the system.

> **Helpful Hint:** Try to eliminate one of the variables from the original first and second
> equations. Then eliminate this same variable from the original second and third
> equations. The result should be a system of two equations in two variables.

$$(1) \quad 3x + 5y - 2z = -5$$
$$(2) \quad 2x + 3y - z = -2$$
$$(3) \quad 2x + 4y + 6z = 18$$

Did you choose the variable z to eliminate?
Yes _____ No _____

If you answered No, consider why this might be the easiest
variable to eliminate out of the three.

Did you multiply equation (2) by -2 and add the result to

equation (1) to obtain the equation $-x - y = -1$?

Yes _____ No _____

If you answered No, stop and perform those operations using
only original equations (2) and (1).

Next did you multiply the original equation (2) by 6 and add

the result to equation (3) to obtain the equation

$14x + 22y = 6$? Yes _____ No _____

If you answered No, stop and perform those
operations using only original equations (2)

and (3).

Make sure that after performing these steps,
your result is a system of two linear equations
in two variables. Once you solve this system
in two variables, remember to go back to one
of the original equations and substitute in the
resulting values for x and y to find the value
of z.

Now go back and rework the problem using
these suggestions.

Copyright © 2013 Pearson Education, Inc.

Solving an Application Using a System of Three Equations in Three Variables—Problem 11

The math club is selling items with the college logo to raise money. Sam bought 4 pens, a mug, and a T-shirt for $20.00. Alicia bought 2 pens and 2 mugs for $11.00. Ramon bought 6 pens, a mug, and 2 T-shirts for $33.00. What was the price of each pen, mug, and T-shirt?

Helpful Hint: Read through the word problem carefully. Specify what each variable represents. Then construct appropriate equations for Sam, Alicia, and Ramon. Solve the resulting system of three linear equations in three variables.

Did you realize that this problem requires a system of three equations in three variables?
Yes _____ No _____

Did you let x = the price of the pens, y = the price of the mugs, and z = the price of the T-shirts?
Yes _____ No _____

If you answered No to either question, go back and read the problem carefully again. Consider that you have three items with unknown prices. Consider that information is provided for three people—Sam, Alicia, and Ramon.

Did you write Sam's equation as $4x + y + z = 20$?
Yes _____ No _____

If you answered No, reread the second sentence in the word problem. Apply the information using the variables listed above. Now translate this information into a linear equation in three variables.

Did you write Alicia's equation as $2x + 2y = 11$? Yes _____ No _____

If you answered No, reread the third sentence in the word problem. Notice that we do not use the variable z because Alicia did not buy any T-shirts.

Now see if you can write a third equation for Ramon that totals $33. Then solve the resulting system.

If you answered Problem 11 incorrectly, go back and rework the problem using these suggestions.

___

Graphing a System of Linear Inequalities—Problem 14

Solve the system of linear inequalities by graphing. $3x + y > 8$
$$x - 2y > 5$$

Helpful Hint: Remember to use dashed lines when the inequality symbols are $>$ or $<$. Take one inequality at a time and graph the border line. Then shade above or below each border line based on test points. The intersection of the two shaded regions is the solution to the system.

First graph the border line $3x + y = 8$. Did you see that this line passes through $(3, -1)$ and $(1, 5)$? Did you see that the line should be dashed?
Yes _____ No _____

If you answered No to any of these questions, stop and examine the first inequality again. Notice the inequality symbol used. Then carefully substitute $x = 3$ into the equation and solve for y. Now you have two ordered pairs to use when graphing the bordered line.

Now you must decide to shade above or below the first border line. Did you substitute $(0,0)$ into $3x + y > 8$ and decide to shade above the line?
Yes _____ No _____

If you answered No, remember that $3(0) + 0$ is not greater than 8. So we do not shade on the side of the line that contains $(0,0)$. We must shade above the line.

Follow this procedure for the second border line. Remember that your solution is the intersection of the two shaded regions.

Now go back and rework the problem using these suggestions.

Copyright © 2013 Pearson Education, Inc.

Chapter 5 Polynomials

5.1 Introduction to Polynomials and Polynomial Functions: Adding, Subtracting, and Multiplying

Vocabulary

polynomial • term • monomial • binomial • trinomial
degree of a term • degree of a polynomial • polynomial in x
descending order • polynomial function • expanding a binomial

1. The _____ is the sum of the exponents of its variables.

2. A(n) _____ is an algebraic expression of one or more terms.

3. The _____ is the degree of the highest-degree term in the polynomial.

4. A(n) _____ is a number, a variable raised to a nonnegative integer power, or a product of numbers and variables raised to nonnegative integer powers.

Example	Student Practice
1. Name the type of polynomial and give its degree.	**2.** Name the type of polynomial and give its degree.
(a) $7x+6$	**(a)** $7x^5 + 2x^4 + 6$
If a variable has no exponent the exponent is understood to be 1. There are two terms and the degree of the highest degree term is 1. This is a binomial of degree 1.	
	(b) $2a^4b^2 + 7a^3b^5 + 11ab$
(b) $5x^2 y + 3xy^3 + 6xy$	
There are three terms and the degree of the highest degree term is $1+3 = 4$. This is a trinomial of degree 4.	**(c)** $8y^3z^7$
(c) $7x^4 y^5$	
This is a monomial of degree 9.	

Vocabulary Answers: 1. degree of a term 2. polynomial 3. degree of a polynomial 4. term

Copyright © 2013 Pearson Education, Inc.

Example	Student Practice
3. Evaluate the polynomial function $p(x) = -3x^3 + 2x^2 - 5x + 6$ to find the following.	**4.** Evaluate the polynomial function $p(x) = -2x^3 + 4x^2 - 7x + 8$ to find the following.

(a) $p(-3)$

$p(-3)$
$= -3(-3)^3 + 2(-3)^2 - 5(-3) + 6$
$= -3(-27) + 2(9) - 5(-3) + 6$
$= 81 + 18 + 15 + 6 = 120$

(b) $p(6)$

$p(6) = -3(6)^3 + 2(6)^2 - 5(6) + 6$
$\quad = -3(216) + 2(36) - 5(6) + 6$
$\quad = -648 + 72 - 30 + 6 = -600$

Student Practice column:

(a) $p(-2)$

(b) $p(4)$

---

5. Add. $(5x^2 - 3x - 8) + (-3x^2 - 7x + 9)$

We remove the parentheses and combine like terms.

$(5x^2 - 3x - 8) + (-3x^2 - 7x + 9)$
$= 5x^2 - 3x - 8 - 3x^2 - 7x + 9$
$= 2x^2 - 10x + 1$

6. Add. $(8y^2 - 2y + 7) + (-6y^2 - 3y + 1)$

---

7. Subtract.
$(-5x^2 - 19x + 15) - (3x^2 - 4x + 13)$

We add the opposite of the second polynomial to the first polynomial.

$(-5x^2 - 19x + 15) - (3x^2 - 4x + 13)$
$= (-5x^2 - 19x + 15) + (-3x^2 + 4x - 13)$
$= -8x^2 - 15x + 2$

8. Subtract.
$(2y^2 + 14y - 9) - (6y^2 - 3y - 5)$

Copyright © 2013 Pearson Education, Inc.

Example	Student Practice
9. Multiply. $(5x+2)(7x-3)$	**10.** Multiply. $(8x-6)(4x+5)$

Using the FOIL method, we multiply the first terms, $5x$ and $7x$, then the outer terms, $5x$ and -3, then the inner terms, 2 and $7x$, and finally the last terms, 2 and -3.

$$(5x+2)(7x-3)$$
$$= \text{First} + \text{Outer} + \text{Inner} + \text{Last}$$
$$= 35x^2 - 15x + 14x - 6$$
$$= 35x^2 - x - 6$$

---

11. Multiply. $(2a-9b)(2a+9b)$

12. Multiply. $(3x-7y)(3x+7y)$

Since the product is of the form $(a+b)(a-b)$, we can use the formula $(a+b)(a-b) = a^2 - b^2$. The product is the difference of two squares.

$$(2a-9b)(2a+9b) = (2a)^2 - (9b)^2$$
$$= 4a^2 - 81b^2$$

We could have used the FOIL method, but recognizing the special product allowed us to save time.

---

13. Multiply. $(5a-8b)^2$

14. Multiply. $(4x+9y^2)^2$

Since the product is the square of a binomial, we can use the formula $(a-b)^2 = a^2 - 2ab + b^2$.

$$(5a-8b)^2 = (5a)^2 - 2(5a)(8b) + (8b)^2$$
$$= 25a^2 - 80ab + 64b^2$$

Copyright © 2013 Pearson Education, Inc.

Example	Student Practice
15. Multiply. $(4x^2 - 2x + 3)(-3x + 4)$	**16.** Multiply. $(3x^2 - 4x + 2)(-5x + 8)$

One way to multiply two polynomials is to write them vertically, as we do when multiplying two- and three-digit numbers. Multiply $(4x^2 - 2x + 3)(+4)$ and multiply $(4x^2 - 2x + 3)(-3x)$. Then add the two products.

$$
\begin{array}{r}
4x^2 - 2x + 3 \\
-3x + 4 \\
\hline
16x^2 - 8x + 12 \\
-12x^3 + 6x^2 - 9x \\
\hline
-12x^3 + 22x^2 - 17x + 12
\end{array}
$$

Extra Practice

1. Subtract. $(-x^2 - 3x + 1) - (5x^2 - 3x + 4)$

2. Multiply. $(2x^2 - 1)(x - 3)$

3. Multiply. $(4x^2 - 5y)^2$

4. Multiply. $(4a - 3)(2a^2 - 5a + 1)$

Concept Check

Explain how you would multiply the following. $(3x + 1)(x - 4)(2x - 3)$

Copyright © 2013 Pearson Education, Inc.

Chapter 5 Polynomials
5.2 Dividing Polynomials

Vocabulary
monomial • polynomial • long division • descending order • coefficient

1. When dividing polynomials, write missing terms with a _____ of zero.

2. When performing long division, both polynomials must be written in _____.

3. When the divisor is a _____, we write the indicated division as the sum of separate fractions and then we reduce each fraction.

4. When we divide polynomials by binomials or trinomials, we perform _____.

Example	**Student Practice**
1. Divide. $\left(15x^3 - 10x^2 + 40x\right) \div 5x$	**2.** Divide. $\left(48y^6 + 32y^4 - 72y^2\right) \div 8y^2$

Example

First we write the indicated division as the sum of separate fractions.

$$\left(15x^3 - 10x^2 + 40x\right) \div 5x$$

$$= \frac{15x^3 - 10x^2 + 40x}{5x}$$

$$= \frac{15x^3}{5x} - \frac{10x^2}{5x} + \frac{40x}{5x}$$

Then, we reduce each fraction, if possible.

$$\frac{15x^3}{5x} - \frac{10x^2}{5x} + \frac{40x}{5x} = 3x^2 - 2x + 8$$

Vocabulary Answers: 1. coefficient 2. descending order 3. monomial 4. long division

Copyright © 2013 Pearson Education, Inc.

Example	Student Practice

3. Divide. $\left(6x^3 + 7x^2 + 3\right) \div (3x - 1)$

4. Divide. $\left(12x^3 + 13x^2 + 6\right) \div (4x - 1)$

We write the division problem in long division form. There is no x-term in the dividend, so we write $0x$. Then we perform long division.

$$
\begin{array}{r}
2x^2 + 3x + 1 \\
3x - 1 \overline{\smash{)}6x^3 + 7x^2 + 0x + 3} \\
\underline{6x^3 - 2x^2} \\
9x^2 + 0x \\
\underline{9x^2 - 3x} \\
3x + 3 \\
\underline{3x - 1} \\
4
\end{array}
$$

The quotient is $2x^2 + 3x + 1$ with a remainder of 4. We may write this as follows:

$$2x^2 + 3x + 1 + \frac{4}{3x - 1}$$

Check the solution.

$$(3x - 1)\left(2x^2 + 3x + 1\right) + 4 \overset{?}{=} 6x^3 + 7x^2 + 3$$

$$6x^3 + 7x^2 - 0x - 1 + 4 \overset{?}{=} 6x^3 + 7x^2 + 3$$

$$6x^3 + 7x^2 + 3 = 6x^3 + 7x^2 + 3$$

Copyright © 2013 Pearson Education, Inc.

Example	Student Practice
5. Divide. $\dfrac{64x^3 - 125}{4x - 5}$	**6.** Divide. $\dfrac{343x^3 - 8}{7x - 2}$

This fraction is equivalent to the problem $\left(64x^3 - 125\right) \div \left(4x - 5\right)$.

Note that two terms are missing in the dividend. We write them with zero coefficients.

$$
\begin{array}{r}
16x^2 + 20x + 25 \\
4x-5\overline{)64x^3 + 0x^2 + 0x - 125} \\
\underline{64x^3 - 80x^2} \\
80x^2 + 0x \\
\underline{80x^2 - 100x} \\
100x - 125 \\
\underline{100x - 125} \\
0
\end{array}
$$

The quotient is $16x^2 + 20x + 25$.

Check the solution.

$$\left(4x - 5\right)\left(16x^2 + 20x + 25\right) \overset{?}{=} 64x^3 - 125$$

$$64x^3 + 0x^2 + 0x - 125 \overset{?}{=} 64x^3 - 125$$

$$64x^3 - 125 = 64x^3 - 125$$

Copyright © 2013 Pearson Education, Inc.

Example	Student Practice

Example

7. Divide.

$$\left(7x^3 - 10x - 7x^2 + 2x^4 + 8\right) \div \left(2x^2 + x - 2\right)$$

Arrange the dividend in descending order before dividing.

$$
\require{enclose}
\begin{array}{r}
x^2 + 3x - 4 \\
2x^2 + x - 2 \enclose{longdiv}{2x^4 + 7x^3 - 7x^2 - 10x + 8} \\
\underline{2x^4 + x^3 - 2x^2} \\
6x^3 - 5x^2 - 10x \\
\underline{6x^3 + 3x^2 - 6x} \\
-8x^2 - 4x + 8 \\
\underline{-8x^2 - 4x + 8} \\
0
\end{array}
$$

The quotient is $x^2 + 3x - 4$.

The check is left to the student.

Student Practice

8. Divide.

$$\left(8x^3 + 8x - 7x^2 + 3x^4 - 4\right) \div \left(x^2 + 3x - 2\right)$$

Extra Practice

1. Divide. $\dfrac{35x^5 y^3 - 14x^4 y^2 - 7x^3 y}{-7x^3 y}$

2. Divide. $\left(x^2 - 6x + 8\right) \div \left(x + 2\right)$

3. Divide. $\left(4x^3 - 5x + 3\right) \div \left(2x - 3\right)$

4. Divide. $\dfrac{2x^4 + x^3 + 14x^2 + 6x + 12}{2x^2 + x + 2}$

Concept Check

Explain how you would check your answer if you divided $\left(x^2 - 9x - 5\right) \div \left(x - 3\right)$ and obtained a result of $x - 6 - \dfrac{23}{x - 3}$.

Copyright © 2013 Pearson Education, Inc.

Name: _____   Date: _____

Instructor: _____   Section: _____

Chapter 5 Polynomials
5.3 Synthetic Division

Vocabulary
synthetic division   •   variables   •   zero   •   degree

1. In synthetic division, we eliminate the _____.

2. The _____ of the quotient should be one less than the degree of the dividend.

3. When dividing a polynomial by a binomial of the form $x + b$ you may find a procedure known as _____ quite efficient.

4. When a term is missing in the sequence of descending powers of x, we use a _____ to indicate the coefficient of that term.

Example	Student Practice
1. Divide using synthetic division. $\left(3x^3 - x^2 + 4x + 8\right) \div \left(x + 2\right)$	**2.** Divide using synthetic division. $\left(5x^3 + 3x^2 - 6x + 10\right) \div \left(x + 3\right)$

First, write the divisor without the 1 and with the opposite sign. Then write the dividend without variables. Next, multiply $(-2)(3) = -6$ and add $-1 + (-6) = -7$.

$$\begin{array}{r|rrrr} -2 & 3 & -1 & 4 & 8 \\ & & -6 & & \\ \hline & 3 & -7 & & \end{array}$$

Follow that process to complete the synthetic division.

$$\begin{array}{r|rrrr} -2 & 3 & -1 & 4 & 8 \\ & & -6 & 14 & -36 \\ \hline & 3 & -7 & 18 & \underline{|-28} \end{array}$$

The quotient is $3x^2 - 7x + 18 + \dfrac{-28}{x + 2}$.

Vocabulary Answers: 1. variables 2. degree 3. synthetic division 4. zero

Copyright © 2013 Pearson Education, Inc.

Example	Student Practice
3. Divide using synthetic division.	**4.** Divide using synthetic division.

3. Divide using synthetic division.

$$\left(3x^4 - 21x^3 + 31x^2 - 25\right) \div \left(x - 5\right)$$

Since $b = -5$, we use 5 as the divisor for synthetic division.

$$
\begin{array}{r|rrrrr}
5 & 3 & -21 & 31 & 0 & -25 \\
  &   & 15 & -30 & 5 & 25 \\
\hline
  & 3 & -6 & 1 & 5 & \underline{|0} \\
\end{array}
$$

Note that the remainder is zero.

The quotient is $3x^3 - 6x^2 + x + 5$.

4. Divide using synthetic division.

$$\left(4x^4 - 29x^3 + 31x^2 - 36\right) \div \left(x - 6\right)$$

---

5. Divide using synthetic division.

$$\left(3x^4 - 4x^3 + 8x^2 - 5x - 5\right) \div \left(x - 2\right)$$

Since $b = -2$, we use 2 as the divisor for synthetic division.

$$
\begin{array}{r|rrrrr}
2 & 3 & -4 & 8 & -5 & -5 \\
  &   & 6 & 4 & 24 & 38 \\
\hline
  & 3 & 2 & 12 & 19 & \underline{|33} \\
\end{array}
$$

The quotient is

$$3x^3 + 2x^2 + 12x + 19 + \frac{33}{x - 2}.$$

6. Divide using synthetic division.

$$\left(5x^4 - 7x^3 + 4x^2 - 3x - 21\right) \div \left(x - 2\right)$$

Copyright © 2013 Pearson Education, Inc.

Extra Practice

1. Divide using synthetic division.

$$(3x^3 - 5x^2 + 7x - 5) \div (x - 1)$$

2. Divide using synthetic division.

$$(3x^4 + x + 2) \div (x - 2)$$

3. Divide using synthetic division.

$$\frac{x^3 - 2x + 6}{x - 4}$$

4. Divide using synthetic division.

$$\frac{2x^5 + 10x^4 + 10x^3 - 4x^2 + x - 4}{x + 2}$$

Concept Check

When doing a problem such as $(2x^4 - x + 3) \div (x - 2),$ why is it necessary to use zeros to represent $0x^3 + 0x^2$ when performing synthetic division?

Copyright © 2013 Pearson Education, Inc.

Copyright © 2013 Pearson Education, Inc.

Chapter 5 Polynomials
5.4 Removing Common Factors; Factoring by Grouping

Vocabulary

factor   •   factoring   •   greatest common factor   •   factoring by grouping

1. Polynomials with four terms can often be factored by the _____ method.

2. The _____ is the largest factor that is common to all terms of the expression.

3. When two or more algebraic expressions are multiplied, each expression is called a _____.

4. _____ is multiplication in reverse.

Example	**Student Practice**
1. Factor out the greatest common factor. $40a^3 - 20a^2$ $40a^3 - 20a^2 = 20a^2(2a - 1)$ The greatest common factor is $20a^2$. Suppose we had written $10a(4a^2 - 2a)$ or $10a(2a)(2a - 1)$ as our answer. Although we have factored the expression, we have not found the greatest common factor.	**2.** Factor out the greatest common factor. $18y - 9y^2$
3. Factor out the greatest common factor. $4a^3 - 12a^2b^2 - 8ab^3 + 6ab$ $4a^3 - 12a^2b^2 - 8ab^3 + 6ab$ $= 2a(2a^2 - 6ab^2 - 4b^3 + 3b)$ The greatest common factor is $2a$.	**4.** Factor out the greatest common factor. $15x^3 - 20x^2y^3 + 40xy^4 - 30xy$

Vocabulary Answers: 1. factoring by grouping 2. greatest common factor 3. factor 4. factoring

Copyright © 2013 Pearson Education, Inc.

Example	Student Practice
5. Factor and check your answer. $6x^3 - 9x^2y - 6x^2y^2$ $6x^3 - 9x^2y - 6x^2y^2 = 3x^2\left(2x - 3y - 2y^2\right)$ Check. First notice that $\left(2x - 3y - 2y^2\right)$ has no common factors. If it did, we would know that we had not factored out the greatest common factor. Next multiply the two factors. $3x^2\left(2x - 3y - 2y^2\right) = 6x^3 - 9x^2y - 6x^2y^2$ We do obtain the original polynomial.	**6.** Factor and check your answer. $24a^3 + 36a^2b - 20a^2b^2$
7. Factor. $5a(a+b) - 2b(a+b) - (a+b)$ $5a(a+b) - 2b(a+b) - (a+b)$ $= 5a(a+b) - 2b(a+b) - 1(a+b)$ $= (a+b)(5a - 2b - 1)$ The common factor is $a+b$.	**8.** Factor. $6(y-4) - 7y(y-4)$
9. Factor. $ax + 2ay + 2bx + 4by$ Remove the greatest common factor (a) from the first two terms. Then remove the greatest common factor $(2b)$ from the last two terms. $ax + 2ay + 2bx + 4by$ $= a(x+2y) + 2b(x+2y)$ Notice that $(x+2y)$ is a common factor. $a(x+2y) + 2b(x+2y)$ $= (x+2y)(a+2b)$	**10.** Factor. $mx + 5nx + 5mz + 25nz$

Copyright © 2013 Pearson Education, Inc.

Example	Student Practice
11. Factor. $2x^2 - 18y - 12x + 3xy$	**12.** Factor. $6x^2 - 20z - 24x + 5xz$

11. Factor. $2x^2 - 18y - 12x + 3xy$

First write the polynomial in this order:
$2x^2 - 12x + 3xy - 18y$

Remove the greatest common factor $(2x)$ from the first two terms. Then remove the greatest common factor $(3y)$ from the last two terms.

$2x^2 - 18y - 12x + 3xy$
$= 2x(x-6) + 3y(x-6)$
$= (x-6)(2x+3y)$

---

13. Factor. $xy - 6 + 3x - 2y$

14. Factor. $mn - 21 + 7n - 3m$

13. Factor. $xy - 6 + 3x - 2y$

Rearrange the terms so that the first two terms have a common factor and the last two terms have a common factor.

$xy - 6 + 3x - 2y = xy + 3x - 2y - 6$

Factor out the common factor x from the first two terms and -2 from the second two terms.

$xy + 3x - 2y - 6 = x(y+3) - 2(y+3)$

Factor out the common binomial factor $y + 3$.

$x(y+3) - 2(y+3) = (y+3)(x-2)$

Copyright © 2013 Pearson Education, Inc.

Example	Student Practice
15. Factor and check your answer by multiplying. $2x^3 + 21 - 7x^2 - 6x$	**16.** Factor and check your answer by multiplying. $4x^3 + 63 - 9x^2 - 28x$

Rearrange the terms. Then factor out a common factor from each group of two terms.

$2x^3 + 21 - 7x^2 - 6x$

$= 2x^3 - 7x^2 - 6x + 21$

$= x^2(2x - 7) - 3(2x - 7)$

Factor out the common binomial factor $2x - 7$.

$x^2(2x - 7) - 3(2x - 7) = (2x - 7)(x^2 - 3)$

Check. Multiply the two binomials.

$(2x - 7)(x^2 - 3) = 2x^3 - 6x - 7x^2 + 21$

$= 2x^3 + 21 - 7x^2 - 6x$

The product is identical to the original expression.

Extra Practice

1. Factor out the greatest common factor.
$30x^3y^3 - 3x^2y^2 + 15xy$

2. Factor out the greatest common factor.
$3x(x - 3) - 4y(x - 3) + 2(x - 3)$

3. Factor. $3x + 4y - 6 - 2xy$

4. Factor. $x^3 - 2x^2 - 3x + 6$

Concept Check

Explain how you would rearrange the order of $8xy - 15 - 10y + 12x$ in order to factor the polynomial.

Copyright © 2013 Pearson Education, Inc.

Name: _____

Date: _____

Instructor: _____

Section: _____

Chapter 5 Polynomials
5.5 Factoring Trinomials

Vocabulary

sum • product • negative • positive • opposite

1. When factoring a trinomial to the form $(x+m)(x+n)$, if the last term of the trinomial is positive and the middle term is negative, the two numbers m and n will be _____ numbers.

2. When factoring a trinomial to the form $(x+m)(x+n)$, if the last term of the trinomial is negative, the two numbers m and n will be _____ in sign.

3. The last term in the trinomial is the _____ of the two numbers m and n when the answer has the form $(x+m)(x+n)$.

4. The coefficient of x in a trinomial is the _____ of the two numbers m and n when the answer has the form $(x+m)(x+n)$.

Example	Student Practice
1. Factor. $x^2 - 14x + 24$	**2.** Factor. $x^2 - 12x + 32$

We want to find two numbers whose product is 24 and whose sum is −14. They will both be negative numbers.

Factor Pairs of 24	Sum of the Factors
$(-24)(-1)$	$-24-1=-25$
$(-12)(-2)$	$-12-2=-14$
$(-8)(-3)$	$-8-3=-11$
$(-6)(-4)$	$-6-4=-10$

The numbers whose product is 24 and whose sum is −14 are −12 and −2. Thus, the factored form is given below.

$$x^2 - 14x + 24 = (x-12)(x-2)$$

Vocabulary Answers: 1. negative 2. opposite 3. product 4. sum

Copyright © 2013 Pearson Education, Inc.

Example	Student Practice
3. Factor. $x^2 + 11x - 26$	**4.** Factor. $x^2 - 9x - 36$

3. Factor. $x^2 + 11x - 26$

We want to find two numbers whose product is -26 and whose sum is 11. One number will be positive and the other negative.

Factor Pairs of -26	Sum of the Factors
$(-26)(1)$	$-26 + 1 = -25$
$(26)(1)$	$26 - 1 = 25$
$(-13)(2)$	$-13 + 2 = -11$
$(13)(-2)$	$13 - 2 = 11$

The numbers whose product is -26 and whose sum is 11 are 13 and -2. Thus, $x^2 + 11x - 26 = (x + 13)(x - 2)$.

4. Factor. $x^2 - 9x - 36$

---

5. Factor. $x^4 - 2x^2 - 24$

We need to recognize that we can write this as $\left(x^2\right)^2 - 2\left(x^2\right) - 24$. We can make this polynomial easier to factor if we substitute y for x^2. Then we have $y^2 + (-2y) + (-24)$. So the factors will be $(y + m)(y + n)$. The two numbers whose product is -24 and whose sum is -2 are -6 and 4. Therefore we have $(y - 6)(y + 4)$. But $y = x^2$, so our answer is as follows.

$x^4 - 2x^2 - 24 = \left(x^2 - 6\right)\left(x^2 + 4\right)$

6. Factor. $x^4 - 3x^2 - 10$

---

7. Factor. $x^2 - 21xy + 20y^2$

$x^2 - 21xy + 20y^2 = (x - 20y)(x - y)$
Notice that the last terms in each factor contain the variable y.

8. Factor. $x^2 + 2xy - 24y^2$

124

Copyright © 2013 Pearson Education, Inc.

Example	Student Practice
9. Factor. $2x^2 + 19x + 24$	**10.** Factor. $3x^2 + 17x + 20$

9. Factor. $2x^2 + 19x + 24$

The grouping number is
$(a)(c) = (2)(24) = 48.$

Now $b = 19$ so we want the factor pair of 48 whose sum is 19. The two factors are 16 and 3. We use the numbers 16 and 3 to write $19x$ as the sum of $16x$ and $3x$.

$2x^2 + 19x + 24 = 2x^2 + 16x + 3x + 24$

Now we factor by grouping.

$2x^2 + 16x + 3x + 24 = 2x(x+8) + 3(x+8)$
$\qquad\qquad\qquad\qquad = (x+8)(2x+3)$

---

11. Factor. $6x^3 - 26x^2 + 24x$

12. Factor. $8x^3 - 36x^2 + 40x$

First we factor out the greatest common factor $2x$ from each term.
$6x^3 - 26x^2 + 24x = 2x(3x^2 - 13x + 12)$

Next we factor $3x^2 - 13x + 12$.

The grouping number is 36. We want the factor pair of 36 whose sum is -13. The two factors are -4 and -9. We write $3x^2 - 13x + 12$ with four terms. Then we factor by grouping. Since we factored out $2x$, it must be part of the answer.

$2x(3x^2 - 13x + 12)$
$= 2x(3x^2 - 4x - 9x + 12)$
$= 2x[x(3x-4) - 3(3x-4)]$
$= 2x(3x-4)(x-3)$

Copyright © 2013 Pearson Education, Inc.

Example	Student Practice
13. Factor by trial and error. $6x^4 + x^2 - 12$	**14.** Factor by trial and error. $$6x^4 - 11x^2 - 10$$

The first term of each factor must contain x^2. Suppose that we try the following possible factors.

$$(2x^2 - 3)(3x^2 + 4)$$

The middle term we get is $-x^2$, but we need its opposite, x^2. In this case, we just need to reverse the signs of -3 and 4. Thus, the factored form is given below.

$$6x^4 + x^2 - 12 = (2x^2 + 3)(3x^2 - 4)$$

Extra Practice

1. Factor. $x^2 - 9x + 14$

2. Factor. $x^2 + 7x - 8$

3. Factor out the greatest common factor from the terms of the trinomial. Then factor the remaining polynomial.
$$2x^2 + 2xy - 12y^2$$

4. Factor. You may use the grouping method or by trial and error.
$$14x^2 - 11x - 15$$

Concept Check

Explain what the first step would be in factoring the following polynomial.
$$3x^2 y^2 + 6xy^2 - 72y^2$$

Copyright © 2013 Pearson Education, Inc.

Chapter 5 Polynomials
5.6 Special Cases of Factoring

Vocabulary
difference of two squares   •   perfect square trinomial   •   sum or difference of two cubes

1. In a _____, the first and last terms are perfect squares and the middle term is twice the product of the values that, when squared, give the first and last terms.

2. When factoring the _____, we use the formula $a^2 - b^2 = (a+b)(a-b)$.

3. To factor a _____, we use the factor formulas $a^3 + b^3 = (a+b)(a^2 - ab + b^2)$ or $a^3 - b^3 = (a-b)(a^2 + ab + b^2)$, respectively.

Example	**Student Practice**
1. Factor. $x^2 - 16$ In this case $a = x$ and $b = 4$ in the formula. $a^2 - b^2 = (a+b) \ (a-b)$ $\downarrow \quad \downarrow \qquad \downarrow \downarrow \quad \downarrow \downarrow$ $(x)^2 - (4)^2 = (x+4) \ (x-4)$	**2.** Factor. $x^2 - 25$
3. Factor. $100w^4 - 9z^4$ In this case $a = 10w^2$ and $b = 3z^2$ in the formula $a^2 - b^2 = (a+b)(a-b)$. $100w^4 - 9z^4 = (10w^2)^2 - (3z^2)^2$ $\qquad = (10w^2 + 3z^2)(10w^2 - 3z^2)$	**4.** Factor. $169x^4 - 16y^4$

Vocabulary Answers: 1. perfect square trinomial 2. difference of two squares 3. sum and difference of two cubes

Copyright © 2013 Pearson Education, Inc.

Example	Student Practice
5. Factor. $25x^2 - 20x + 4$	**6.** Factor. $9x^2 - 24x + 16$

5. Factor. $25x^2 - 20x + 4$

This trinomial is a perfect square. The first and last terms are perfect squares.

$$25x^2 - 20x + 4 = (5x)^2 - 20x + (2)^2$$

The middle term is twice the product of the value $5x$ and the value 2. In other words, $2(5x)(2) = 20x$. So, we can use the formula $a^2 - 2ab + b^2 = (a-b)^2$.

$$25x^2 - 20x + 4 = (5x)^2 - 2(5x)(2) + (2)^2$$
$$= (5x - 2)^2$$

---

7. Factor. $200x^2 + 360x + 162$ **8.** Factor. $242x^2 + 308x + 98$

First we factor out the common factor 2. Then we can use the formula
$a^2 + 2ab + b^2 = (a+b)^2$.

$$200x^2 + 360x + 162$$
$$= 2(100x^2 + 180x + 81)$$
$$= 2\left[(10x)^2 + (2)(10x)(9) + (9)^2\right]$$
$$= 2(10x + 9)^2$$

---

9. Factor. $9x^4 + 30x^2y^2 + 25y^4$ **10.** Factor. $36x^4 + 84x^2y^2 + 49y^4$

$$9x^4 + 30x^2y^2 + 25y^4$$
$$= (3x^2)^2 + 2(3x^2)(5y^2) + (5y^2)^2$$
$$= (3x^2 + 5y^2)^2$$

Copyright © 2013 Pearson Education, Inc.

Example	Student Practice
11. Factor. $125x^3 + y^3$ Here $a = 5x$ and $b = y$. We can use the formula $a^3 + b^3 = (a+b)(a^2 - ab + b^2)$. $125x^3 + y^3 = (5x)^3 + (y)^3$ $\qquad = (5x+y)(25x^2 - 5xy + y^2)$	**12.** Factor. $64x^3 + 27y^3$
13. Factor. $125w^3 - 8z^6$ Here $a = 4x$ and $b = 2z^2$. We can use the difference of two cubes formula, $a^3 - b^3 = (a-b)(a^2 + ab + b^2)$. $125w^3 - 8z^6$ $= (5w)^3 - (2z^2)^3$ $= (5w - 2z^2)(25w^2 + 10wz^2 + 4z^4)$	**14.** Factor. $216x^6 - 343y^9$
15. Factor. $250x^3 - 2$ First we factor out the common factor of 2. $250x^3 - 2 = 2(125x^3 - 1)$ Now we can use the formula for the difference of two cubes, $a^3 - b^3 = (a-b)(a^2 + ab + b^2)$. $250x^3 - 2 = 2(125x^3 - 1)$ $\qquad = 2(5x - 1)(25x^2 + 5x + 1)$ Note that the trinomial cannot be factored.	**16.** Factor. $192x^3 - 3$

Copyright © 2013 Pearson Education, Inc.

1. Factor. $18x^3 - 32xy^2$

2. Factor. $4x^2 - 12x + 9$

3. Factor. $48x^2 - 144x + 108$

4. Factor. $81x^3 - 24y^3$

Concept Check

Explain why the formula $(a+b)^2 = a^2 + 2ab + b^2$ does not work when factoring the following expression. $36x^2 + 66xy + 121y^2$

Copyright © 2013 Pearson Education, Inc.

Name: _____   Date: _____
Instructor: _____   Section: _____

Chapter 5 Polynomials
5.7 Factoring a Polynomial Completely

Vocabulary
greatest common factor   •   difference of two squares   •   difference of two cubes
sum of two cubes   •   perfect square trinomial   •   factor by grouping   •   prime

1. If a polynomial cannot be factored using rational numbers, it is said to be _____.

2. When factoring a polynomial, factor out the _____ before doing anything else.

3. If a polynomial has four terms, try to _____.

4. If a polynomial has three terms and the form $a^2 + 2ab + b^2$ it is a _____.

Example	**Student Practice**
1. Factor completely.	**2.** Factor completely.
(a) $2x^2 - 18$	**(a)** $8x^2 - 50$
Factor out the common factor. Then use $a^2 - b^2 = (a+b)(a-b)$.	
$\begin{aligned} 2x^2 - 18 &= 2(x^2 - 9) \\ &= 2(x+3)(x-3) \end{aligned}$	
(b) $27x^4 - 8x$	**(b)** $16x^5 - 54x^2$
Factor out the common factor. Then use $a^3 - b^3 = (a-b)(a^2 + ab + b^2)$.	
$\begin{aligned} 27x^4 - 8x &= x(27x^3 - 8) \\ &= x(3x-2)(9x^2 + 6x + 4) \end{aligned}$	

Vocabulary Answers: 1. prime 2. greatest common factor 3. factor by grouping 4. perfect square trinomial

Copyright © 2013 Pearson Education, Inc.

Example	Student Practice
3. Factor completely.	**4.** Factor completely.

3. Factor completely.

(a) $27x^2 + 36xy + 12y^2$

Factor out the common factor. Then use $(a+b)^2 = a^2 + 2ab + b^2$.

$27x^2 + 36xy + 12y^2$
$= 3(9x^2 + 12xy + 4y^2) = 3(3x+2y)^2$

(b) $2x^2 - 100x + 98$

Factor out the common factor. Then the trinomial will have the form $x^2 + bx + c$.

$2x^2 - 100x + 98 = 2(x^2 - 50x + 49)$
$= 2(x-49)(x-1)$

(c) $6x^3 + 11x^2 - 10x$

Factor out the common factor. Then the trinomial will have the form $ax^2 + bx + c$.

$6x^3 + 11x^2 - 10x = x(6x^2 + 11x - 10)$
$= x(3x-2)(2x+5)$

(d) $5ax + 5ay - 20x - 20y$

Factor out the common factor. Then factor by grouping.

$5ax + 5ay - 20x - 20y$
$= 5(ax + ay - 4x - 4y)$
$= 5[a(x+y) - 4(x+y)]$
$= 5(x+y)(a-4)$

4. Factor completely.

(a) $100x^2 + 120xy + 36y^2$

(b) $3x^2 - 45x + 108$

(c) $12x^4 - x^3 - 6x^2$

(d) $6bw - 6bz + 42w - 42z$

Copyright © 2013 Pearson Education, Inc.

Example	Student Practice
5. If possible, factor $6x^2 + 10x + 3$.	**6.** If possible, factor $7x^2 + 12x + 4$.

5. The trinomial has the form $ax^2 + bx + c$. The grouping number is 18. If the trinomial can be factored, we must find two numbers whose product is 18 and whose sum is 10.

Factor Pairs of 18	Sum of the Factors
$(18)(1)$	19
$(9)(2)$	11
$(6)(3)$	9

There are no numbers meeting the necessary conditions. Thus, the polynomial is prime. (If you use the trial-and-error method, try all possible factors and show that none of them has a product with a middle term of $10x$.

7. If possible, factor $25x^2 + 49$.	**8.** If possible, factor $64x^2 + 81y^2$.

7. Unless there is a common factor that can be factored out, binomials of the form $a^2 + b^2$ cannot be factored. Therefore, $25x^2 + 49$ is prime.

Copyright © 2013 Pearson Education, Inc.

Extra Practice

1. If possible, factor $x^2 + 9$.

2. If possible, factor $2x^3 - 24x^2 + 72x$.

3. If possible, factor $x^4 - 1$.

4. If possible, factor $x^3 + 3x^2 - 4x - 12$.

Concept Check

Explain why $4x^2 + 3x + 1$ is prime.

Copyright © 2013 Pearson Education, Inc.

Name: _____   Date: _____

Instructor: _____   Section: _____

Chapter 5 Polynomials
5.8 Solving Equations and Applications Using Polynomials

Vocabulary
quadratic equation   •   standard form   •   zero factor property   •   double root

1. The _____ states that for all real numbers a and b, if $a \cdot b = 0$, then $a = 0$, $b = 0$, or both $= 0$.

2. A second-degree equation of the form $ax^2 + bx + c = 0$, where a, b, and c are real numbers and $a \neq 0$, is a _____.

3. One solution that is obtained twice is called a _____.

Example	Student Practice
1. Solve the equation. $x^2 + 15x = 100$	**2.** Solve the equation. $x^2 + 13x = 30$

When we say "solve the equation" or "find the roots," we mean "find the values of x that satisfy the equation." First subtract 100 from both sides. Then factor the trinomial. Set each factor equal to 0 and solve each equation.
$$x^2 + 15x = 100$$
$$x^2 + 15x - 100 = 0$$
$$(x+20)(x-5) = 0$$
$$x + 20 = 0 \quad \text{or} \quad x - 5 = 0$$
$$x = -20 \qquad x = 5$$

Use the original equation to check.
$$x = -20: \ (-20)^2 + 15(-20) \overset{?}{=} 100$$
$$100 = 100$$
$$x = 5: \ (5)^2 + 15(5) \overset{?}{=} 100$$
$$100 = 100$$

Thus, 5 and -20 are both roots.

Vocabulary Answers: 1. zero factor property 2. quadratic equation 3. double root

Copyright © 2013 Pearson Education, Inc.

Example	Student Practice
3. Solve. $9x(x-1)=3x-4$	**4.** Solve. $5x(5x-4)=20x-16$

3. Solve. $9x(x-1)=3x-4$

First remove parentheses. Then get 0 on one side and combine like terms. Then factor.

$$9x(x-1)=3x-4$$
$$9x^2-9x-3x+4=0$$
$$9x^2-12x+4=0$$
$$(3x-2)^2=0$$
$$3x-2=0 \quad \text{or} \quad 3x-2=0$$
$$3x=2 \qquad\qquad 3x=2$$
$$x=\frac{2}{3} \qquad\qquad x=\frac{2}{3}$$

We obtain one solution twice. This value is called a double root.

5. Solve. $2x^3=24x-8x^2$

First get 0 on one side of the equation. Then factor out the common factor $2x$.

$$2x^3=24x-8x^2$$
$$2x^3+8x^2-24x=0$$
$$2x(x^2+4x-12)=0$$

Factor the trinomial and solve for x.

$$2x(x^2+4x-12)=0$$
$$2x(x+6)(x-2)=0$$
$$2x=0 \quad \text{or} \quad x+6=0 \quad \text{or} \quad x-2=0$$
$$x=0 \qquad\qquad x=-6 \qquad\qquad x=2$$

The solutions are 0, −6, and 2.

6. Solve. $3x^3=84x+9x^2$

Copyright © 2013 Pearson Education, Inc.

Example	Student Practice
7. A racing sailboat has a triangular sail. Find the base and altitude of the triangular sail that has an area of 35 square meters and a base that is 3 meters shorter that the altitude.	**8.** A racing sailboat has a triangular sail. Find the base and altitude of the triangular sail that has an area of 30 square meters and a base that is 7 meters shorter that the altitude.

To understand the problem, draw a sketch and recall the formula for the area of a triangle, $\text{Area} = \dfrac{1}{2}ab$ where $a = \text{altitude}$ and $b = \text{base}$.

Let $x = $ the length of the altitude in meters. Then $x - 3 = $ the length of the base in meters.

Write an equation. Replace A with 35, a with x, and b with $x - 3$. Then solve the equation.

$$A = \frac{1}{2}ab$$

$$35 = \frac{1}{2}x(x-3)$$

$$70 = x(x-3)$$

$$70 = x^2 - 3x$$

$$0 = x^2 - 3x - 70$$

$$0 = (x-10)(x+7)$$

$$x = 10 \quad \text{or} \quad x = -7$$

The altitude of a triangle must be a positive number, so we disregard -7.

$\text{altitude} = x = 10$ meters

$\text{base} = x - 3 = 10 - 3 = 7$ meters

The altitude of the triangular sail measures 10 meters, and the base of the sail measures 7 meters.

Copyright © 2013 Pearson Education, Inc.

Extra Practice

1. Find the roots and check your answers.

   $x^2 + 16 = 10x$

2. Find the roots and check your answers.

   $x(7x - 3) + 10 = -5(x - 2)$

3. The area of the base of a rectangular desk telephone is 70 square centimeters. The length of the rectangular desk telephone is 3 centimeters longer than its width. What are the length and width of the desk telephone?

4. A picture frame company adds a new square picture frame to its collection each year. This year's frame has an area that is 32 square centimeters greater than the area of the square picture frame introduced last year. The length of each side of the new frame is 5 centimeters less than double the length of last year's frame. Find the dimensions of this year's frame.

Concept Check

Explain how you would solve the following equation. $\dfrac{3}{8}x^2 + \dfrac{1}{4}x = 1$

Copyright © 2013 Pearson Education, Inc.

MATH COACH

Mastering the skills you need to do well on the test.

Watch the **MATH COACH** videos in MyMathLab® or on You Tube while you work the problems below. These helpful hints will help you avoid making common errors on test problems.

Performing Long Division with Polynomials—Problem 7

Divide. $\left(2x^4 - 7x^3 + 7x^2 - 9x + 10\right) \div (2x - 5)$

Helpful Hint: Be careful when multiplying the first term of the quotient by the divisor. Make sure to multiply by both terms of the divisor. Watch out for $+$ and $-$ signs during the subtraction process. This is where most errors occur.

After the first step of division did you obtain x^3 for the first term of the quotient? Yes _____ No _____

When you multiplied x^3 by $2x - 5$, did you get $2x^4 - 5x^3$? Yes _____ No _____

When you subtracted for the first time, did you get $-2x^3$? Yes _____ No _____

If you answered No to any of these questions, go back and complete these steps again. Remember that
$\left(2x^4 - 7x^3\right) - \left(2x^4 - 5x^3\right) = 2x^4 - 7x^3 - 2x^4 + 5x^3.$

After the second step of division, did you obtain $-x^2$ for the second term of the quotient? Yes _____ No _____

When you multiplied $-x^2$ by $2x - 5$, did you get $-2x^3 + 5x^2$? Yes _____ No _____

When you subtracted for the second time, did you get $2x^2$? Yes _____ No _____

If you answered No to any of these questions, go back and complete these steps again.

Remember that $\left(-2x^3 + 7x^2\right) - \left(2x^3 - 5x^2\right)$
$$= -2x^3 + 7x^2 + 2x^3 - 5x^2.$$

If you answered Problem 7 incorrectly, go back and rework the problem using these suggestions.

Factoring a Trinomial with a Greatest Common Factor—Problem 16

Factor, if possible. $18x^2 + 3x - 15$

Helpful Hint: Always look for a common factor before performing any other factoring step. Factor out the greatest common factor first. Then try to factor the remaining trinomial.

Did you notice that the greatest common factor of the trinomial is 3? Yes _____ No _____

After factoring out the GCF, did you obtain $3\left(6x^2 + x - 5\right)$? Yes _____ No _____

If you answered No to either question, carefully perform that step again. Remember to factor out the GCF of 3 from each term.

Did you notice that $6x^2 + x - 5$ can be factored into the form $(?x \ 1)(?x \ 5)$? Yes _____ No _____

Did you notice that the parentheses must end with one positive number and with one negative number since the product must be -5? Yes _____ No _____

If you answered No, stop and complete this step again.

Now go back and rework the problem using these suggestions.

Copyright © 2013 Pearson Education, Inc.

Factoring a Polynomial with Four Terms—Problem 20   Factor, if possible. $3x - 10ay + 6y - 5ax$

> **Helpful Hint:** Check to see if the first two terms have a common factor. If they do not, rearrange the order of the terms of the polynomial. Group the terms so that the first two terms have a common factor and the last two terms have a common factor.

Did you rearrange the order of the terms of the polynomial to $3x + 6y - 10ay - 5ax$?

Yes _____   No _____

Did you factor out a common factor of 3 from the first two terms to obtain $3(x + 2y)$?

Yes _____   No _____

If you answered No to these questions, go back and interchange the second term with the third term. Then factor out a common factor of 3 from the first two terms of the rearranged polynomial.

Did you factor out a common factor of $-5a$ from the second two terms of the rearranged polynomial to obtain $-5a(2y + x)$?

Yes _____   No _____

If you answered No, go back to the rearranged polynomial and factor out the GCF of the last two terms. Note that by the commutative property of addition, the expression $(2y + x)$ is equivalent to $(x + 2y)$. Simplify your result so that the common factor appears only once.

If you answered Problem 20 incorrectly, go back and rework the problem using these suggestions.

Solving a Quadratic Equation by Factoring—Problem 23   Solve. $x^2 = 5x + 14$

> **Helpful Hint:** Before factoring, write the equation in standard form, which is $ax^2 + bx + c = 0$.

Did you first add $-5x - 14$ to each side of the equation to obtain $x^2 - 5x - 14 = 0$?

Yes _____   No _____

If you answered No, stop and perform that step so that your equation is written in standard form.

Did you factor the rewritten equation to obtain $(x - 7)(x + 2) = 0$ or $(x + 2)(x - 7) = 0$?

Yes _____   No _____

If you answered No, go back and complete the factoring step again. Be careful of signs. Remember that once you finish factoring, you must set each factor equal to 0 to find the two solutions to the equation.

Now go back and rework the problem using these suggestions.

Copyright © 2013 Pearson Education, Inc.

Chapter 6 Rational Expressions and Equations
6.1 Rational Expressions and Functions: Simplifying, Multiplying, and Dividing

Vocabulary
rational number   •   rational expression   •   rational function   •   domain
basic rule of fractions   •   common factor   •   reciprocal

1. A function defined by a rational expression is a _____.

2. To divide two rational expressions, take the _____ of the second expression and then multiply.

3. When you divide out a common factor from the numerator and the denominator of a fraction, you are applying the _____.

4. A _____ is of the form $\dfrac{P}{Q}$, where P and Q are polynomials and Q is not zero.

Example	Student Practice
1. Find the domain of the function $f(x) = \dfrac{x-7}{x^2+8x-20}$.	**2.** Find the domain of the function $f(x) = \dfrac{7x+1}{x^2+12x-45}$.

The domain of $f(x)$ will be all real numbers except those that make the denominator equal to zero. Find these value(s) by solving the equation $x^2+8x-20=0$. First, factor the left side of the equation, $(x+10)(x-2)=0$.

Now use the zero factor property and solve for x.

$$x+10=0 \quad \text{or} \quad x-2=0$$
$$x=-10 \qquad\qquad x=2$$

The domain of $y=f(x)$ is all real numbers except -10 and 2.

Vocabulary Answers: 1. rational function 2. reciprocal 3. basic rule of fractions 4. rational expression

Copyright © 2013 Pearson Education, Inc.

Example	Student Practice
3. Simplify. $\dfrac{2a^2 - ab - b^2}{a^2 - b^2}$	**4.** Simplify. $\dfrac{144m^2 - n^2}{12m^2 + 13mn + n^2}$

Begin by factoring the numerator and the denominator. Then apply the basic rule of fractions.

$$\frac{2a^2 - ab - b^2}{a^2 - b^2} = \frac{(2a+b)(a-b)}{(a+b)(a-b)}$$

$$= \frac{2a+b}{a+b} \cdot 1 = \frac{2a+b}{a+b}$$

As you become more familiar with the basic rule, you won't have to write out every step. We did so here to show the application of the rule. We cannot simplify this fraction any further.

Example	Student Practice
5. Simplify. $\dfrac{25y^2 - 16x^2}{8x^2 - 14xy + 5y^2}$	**6.** Simplify. $\dfrac{27y^2 - 75xy + 8x^2}{64x^2 - 9y^2}$

Begin by factoring the numerator and the denominator. Observe that $4x - 5y = -1(-4x + 5y)$.

$$\frac{(5y+4x)(5y-4x)}{(4x-5y)(2x-y)}$$

$$= \frac{(5y+4x)(5y-4x)}{-1(-4x+5y)(2x-y)}$$

Apply the basic rule of fractions and simplify.

$$\frac{(5y+4x)(5y-4x)}{-1(-4x+5y)(2x-y)} = \frac{5y+4x}{-1(2x-y)}$$

$$= -\frac{5y+4x}{2x-y}$$

Copyright © 2013 Pearson Education, Inc.

Example	Student Practice
7. Multiply.	**8.** Multiply.

7. Multiply.

$$\frac{2x^2 - 4x}{x^2 - 5x + 6} \cdot \frac{x^2 - 9}{2x^4 + 14x^3 + 24x^2}$$

We first use the basic rule of fractions; that is, we factor (if possible) the numerator and denominator and divide out common factors.

$$\frac{2x(x-2)}{(x-2)(x-3)} \cdot \frac{(x+3)(x-3)}{2x^2(x^2 + 7x + 12)}$$

$$= \frac{2x(x-2)(x+3)(x-3)}{(2x)x(x-2)(x-3)(x+3)(x+4)}$$

$$= \frac{2x}{2x} \cdot \frac{1}{x} \cdot \frac{x-2}{x-2} \cdot \frac{x+3}{x+3} \cdot \frac{x-3}{x-3} \cdot \frac{1}{x+4}$$

$$= 1 \cdot \frac{1}{x} \cdot 1 \cdot 1 \cdot 1 \cdot \frac{1}{x+4}$$

$$= \frac{1}{x(x+4)} \quad \text{or} \quad \frac{1}{x^2 + 4x}$$

Although either answer form is correct, we usually use the factored form.

8. Multiply.

$$\frac{3x^2 - 15x}{x^2 - 9x + 20} \cdot \frac{x^2 - 16}{6x^4 + 66x^3 + 168x^2}$$

---

9. Multiply. $\dfrac{7x + 7y}{4ax + 4ay} \cdot \dfrac{8a^2x^2 - 8b^2x^2}{35ax^3 - 35bx^3}$

Use the basic rule of fractions and simplify. Note that we can shorten our steps by not writing every factor 1.

$$\frac{7(x+y)}{4a(x+y)} \cdot \frac{8x^2(a+b)(a-b)}{35x^3(a-b)}$$

$$= \frac{\cancel{7}(\cancel{x+y})}{\cancel{4}a(\cancel{x+y})} \cdot \frac{\overset{2}{\cancel{8}}\,\cancel{x^2}(a+b)\cancel{(a-b)}}{\underset{5}{\cancel{35}}\,\underset{x}{\cancel{x^3}}\cancel{(a-b)}}$$

$$= \frac{2(a+b)}{5ax} \quad \text{or} \quad \frac{2a + 2b}{5ax}$$

10. Multiply. $\dfrac{11x + 11y}{3ax + 3ay} \cdot \dfrac{15a^2x - 15b^2x}{99ax^4 - 99bx^4}$

Copyright © 2013 Pearson Education, Inc.

Example	Student Practice

Example

11. Divide. $\dfrac{4x^2 - y^2}{x^2 + 4xy + 4y^2} \div \dfrac{4x - 2y}{3x + 6y}$

We take the reciprocal of the second fraction and multiply the fractions.

$$\dfrac{4x^2 - y^2}{x^2 + 4xy + 4y^2} \cdot \dfrac{3x + 6y}{4x - 2y}$$

$$= \dfrac{(2x + y)(2x - y)}{(x + 2y)(x + 2y)} \cdot \dfrac{3(x + 2y)}{2(2x - y)}$$

$$= \dfrac{3(2x + y)}{2(x + 2y)} \quad \text{or} \quad \dfrac{6x + 3y}{2x + 4y}$$

Student Practice

12. Divide.

$$\dfrac{36a^2 - b^2}{a^2 + 20ab + 100b^2} \div \dfrac{42ax - 7bx}{2a + 20b}$$

Extra Practice

1. Simplify completely. $\dfrac{x^2 - 4}{6x - 12}$

2. Simplify completely. $\dfrac{-3y - 18y^2}{12y^2 - 16y - 3}$

3. Multiply. $\dfrac{3x - 3y}{3x^2 - 5xy + 2y^2} \cdot \dfrac{3x^3 - 2x^2 y}{3x^2 + 12}$

4. Divide. $\dfrac{x^2 + 6x + 8}{2x^2 + 9x + 4} \div (x + 2)$

Concept Check

Explain how you would simplify the following. $\dfrac{9 - x^2}{x^2 - 7x + 12}$

Copyright © 2013 Pearson Education, Inc.

Chapter 6 Rational Expressions and Equations
6.2 Adding and Subtracting Rational Expressions

Vocabulary
rational expression   •   numerator   •   denominator   •   prime factor
least common denominator (LCD)   •   equivalent fraction

1. The _____ of two rational expressions is the product of the different prime factors in each denominator. If a factor is repeated, we use the highest power that appears on the factor in the denominators.

2. To add or subtract rational expressions, the _____ of each fraction must be the same.

3. If two rational expressions have different denominators, rewrite each fraction as a(n) _____ that has the LCD as the denominator.

4. When two rational expressions are subtracted, be careful with the signs in the _____ of the second fraction.

Example	**Student Practice**
1. Find the LCD of the rational expressions $\dfrac{7}{x^2-4}$ and $\dfrac{2}{x-2}$.	**2.** Find the LCD of the rational expressions $\dfrac{22}{49x^2-9y^2}$ and $\dfrac{6}{7x+3y}$.

First we factor each denominator completely (into prime factors).

$$x^2 - 4 = (x+2)(x-2)$$

$x - 2$ cannot be factored.

Next we list all the different prime factors. The different factors are $x+2$ and $x-2$.

Since no factor occurs more than once in the denominators, the LCD is the product $(x+2)(x-2)$.

Vocabulary Answers: 1. least common denominator (LCD) 2. denominator 3. equivalent fraction
4. numerator

Copyright © 2013 Pearson Education, Inc.

Example	Student Practice

3. Find the LCD of the rational expressions $\dfrac{7}{12xy^2}$ and $\dfrac{4}{15x^3y}$.

We factor each denominator.

$12xy^2 = 2 \cdot 2 \cdot 3 \cdot x \cdot y \cdot y$

$15x^3y = 3 \cdot 5 \cdot x \cdot x \cdot x \cdot y$

Our LCD will require each of the different factors that appear in either denominator. They are 2, 3, 5, x, and y.

The factor 2 and the factor y each occur twice in $12xy^2$. The factor x occurs three times in $15x^3y$. Thus, the LCD is as follows.

$LCD = 2 \cdot 2 \cdot 3 \cdot 5 \cdot x \cdot x \cdot x \cdot y \cdot y$

$\quad\quad = 60x^3y^2$

4. Find the LCD of the rational expressions $\dfrac{3}{8x^4y^3}$ and $\dfrac{19}{20x^2y^2}$.

---

5. Subtract.

$$\frac{5x+2}{(x+3)(x-4)} - \frac{6x}{(x+3)(x-4)}$$

We can add or subtract rational expressions with the same denominator just as we do in arithmetic: We simply add or subtract the numerators.

$$\frac{5x+2}{(x+3)(x-4)} - \frac{6x}{(x+3)(x-4)}$$

$$= \frac{5x+2-6x}{(x+3)(x-4)}$$

$$= \frac{-x+2}{(x+3)(x-4)}$$

6. Subtract.

$$\frac{12x-1}{(3x+5)(x-8)} - \frac{7x}{(3x+5)(x-8)}$$

Copyright © 2013 Pearson Education, Inc.

Example	Student Practice
7. Add the rational expressions. $$\dfrac{7}{(x+2)(x-2)}+\dfrac{2}{x-2}$$ The two rational expressions have different denominators. Before we can add these fractions, we must rewrite our fractions with the LCD. The LCD is $(x+2)(x-2)$. The first fraction needs no change, but the second fraction does. $$\dfrac{7}{(x+2)(x-2)}+\dfrac{2}{x-2}\cdot\dfrac{x+2}{x+2}$$ Since $\dfrac{x+2}{x+2}=1,$ we have not changed the value of the fraction. We are simply writing it in an equivalent form. Now add the fractions. $$\dfrac{7}{(x+2)(x-2)}+\dfrac{2(x+2)}{(x+2)(x-2)}$$ $$=\dfrac{7+2(x+2)}{(x+2)(x-2)}$$ $$=\dfrac{7+2x+4}{(x+2)(x-2)}=\dfrac{2x+11}{(x+2)(x-2)}$$	**8.** Add the rational expressions. $$\dfrac{14}{x+9}+\dfrac{5}{(x+9)(x-1)}$$
9. Add. $\dfrac{7}{2x^2y}+\dfrac{3}{xy^2}$ You should be able to see that the LCD of these fractions is $2x^2y^2$. Write equivalent fractions and then add. $$\dfrac{7}{2x^2y}\cdot\dfrac{y}{y}+\dfrac{3}{xy^2}\cdot\dfrac{2x}{2x}=\dfrac{7y}{2x^2y^2}+\dfrac{6x}{2x^2y^2}$$ $$=\dfrac{6x+7y}{2x^2y^2}$$	**10.** Add. $\dfrac{1}{5x^4y}+\dfrac{3}{8x^2y^3}$

147

Copyright © 2013 Pearson Education, Inc.

Example	Student Practice
11. Subtract. $\dfrac{2x+1}{25x^2+10x+1}-\dfrac{6x}{25x+5}$	**12.** Subtract. $\dfrac{8x-1}{9x^2+42x+49}-\dfrac{2x}{9x+21}$

First factor each denominator into prime factors, $25x^2+10x+1=(5x+1)^2$ and $25x+5=5(5x+1)$. To find the LCD, use the highest power of each different factor. Thus, the LCD is $5(5x+1)^2$.

Write our problem in factored form. Then rewrite the fractions as equivalent fractions with the LCD and subtract.

$$\dfrac{2x+1}{(5x+1)^2}-\dfrac{6x}{5(5x+1)}$$

$$=\dfrac{2x+1}{(5x+1)^2}\cdot\dfrac{5}{5}-\dfrac{6x}{5(5x+1)}\cdot\dfrac{5x+1}{5x+1}$$

$$=\dfrac{5(2x+1)-6x(5x+1)}{5(5x+1)^2}$$

$$=\dfrac{-30x^2+4x+5}{5(5x+1)^2}$$

Extra Practice

1. Find the LCD.

$$\dfrac{5}{3x^2-x},\ \dfrac{8}{3x^2+5x-2}$$

2. Subtract and simplify your answer.

$$\dfrac{3x+1}{x^2-11x+6}-\dfrac{7x}{x^2-11x+6}$$

3. Add and simplify your answer.

$$\dfrac{4}{y^2-3y+2}+\dfrac{5}{y^2-1}$$

4. Subtract and simplify your answer.

$$\dfrac{5x}{5x^2-17x+6}-\dfrac{4}{5x-2}$$

Concept Check

Explain how you would find the LCD of $\dfrac{2}{x^2-16}$ and $\dfrac{3}{x^2-6x+8}$.

Copyright © 2013 Pearson Education, Inc.

Name: _____ Date: _____

Instructor: _____ Section: _____

Chapter 6 Rational Expressions and Equations
6.3 Complex Rational Expressions

Vocabulary
complex rational expression • complex fraction • numerator • denominator • LCD

1. A(n) _____ is a large fraction that has at least one rational expression in the numerator, in the denominator, or in both the numerator and the denominator.

2. Method 1 of simplifying complex rational expressions first combines fractions in both the _____ and denominator.

3. A complex rational expression may also be called a(n) _____.

4. Method 2 of simplifying complex rational expressions involves multiplying the numerator and denominator by the _____ of all individual fractions.

Example	Student Practice
1. Simplify. $\dfrac{x+\dfrac{1}{x}}{\dfrac{1}{x}+\dfrac{3}{x^2}}$	2. Simplify. $\dfrac{m+\dfrac{8}{m}}{\dfrac{4}{m^3}+\dfrac{2}{m^2}}$

First find the LCD of all the fractions in the numerator and denominator. The LCD is x^2. Multiply the numerator and denominator by the LCD. Use the distributive property.

$$\dfrac{x+\dfrac{1}{x}}{\dfrac{1}{x}+\dfrac{3}{x^2}} \cdot \dfrac{x^2}{x^2} = \dfrac{x^3+x}{x+3}$$

The result is already simplified, but we will write it in factored form.

$$\dfrac{x^3+x}{x+3} = \dfrac{x\left(x^2+1\right)}{x+3}$$

Vocabulary Answers: 1. complex rational expression 2. numerator 3. complex fraction 4. LCD

Copyright © 2013 Pearson Education, Inc.

Example	Student Practice

Example

3. Simplify. $\dfrac{\dfrac{1}{2x+6}+\dfrac{3}{2}}{\dfrac{3}{x^2-9}+\dfrac{x}{x-3}}$

Student Practice

4. Simplify. $\dfrac{\dfrac{5}{7}+\dfrac{1}{7y-35}}{\dfrac{1}{y^2-25}+\dfrac{y}{y+5}}$

First simplify the numerator.

$$\frac{1}{2x+6}+\frac{3}{2}=\frac{1}{2(x+3)}+\frac{3}{2}$$

$$=\frac{1}{2(x+3)}+\frac{3(x+3)}{2(x+3)}$$

$$=\frac{3x+10}{2(x+3)}$$

Simplify the denominator.

$$\frac{3}{x^2-9}+\frac{x}{x-3}$$

$$=\frac{3}{(x+3)(x-3)}+\frac{x}{x-3}$$

$$=\frac{3}{(x+3)(x-3)}+\frac{x(x+3)}{(x+3)(x-3)}$$

$$=\frac{x^2+3x+3}{(x+3)(x-3)}$$

Now divide the numerator by the denominator.

$$\frac{3x+10}{2(x+3)}\div\frac{x^2+3x+3}{(x+3)(x-3)}$$

$$=\frac{3x+10}{2\cancel{(x+3)}}\cdot\frac{\cancel{(x+3)}(x-3)}{x^2+3x+3}$$

$$=\frac{(3x+10)(x-3)}{2(x^2+3x+3)}$$

The answer is fully simplified.

Copyright © 2013 Pearson Education, Inc.

Example	Student Practice
5. Simplify by Method 1: Combining fractions in both numerator and denominator.	**6.** Simplify by Method 1: Combining fractions in both numerator and denominator.

5. Simplify by Method 1: Combining fractions in both numerator and denominator.

$$\frac{x+3}{\dfrac{9}{x}-x}$$

Notice that the numerator is fully simplified. Now simplify the denominator.

$$\frac{x+3}{\dfrac{9}{x}-\dfrac{x}{1}\cdot\dfrac{x}{x}} = \frac{x+3}{\dfrac{9}{x}-\dfrac{x^2}{x}}$$

$$= \frac{x+3}{\dfrac{9-x^2}{x}}$$

Write the numerator as a fraction with denominator 1.

$$\frac{x+3}{\dfrac{9-x^2}{x}} = \frac{\dfrac{x+3}{1}}{\dfrac{9-x^2}{x}}$$

Now divide the numerator by the denominator.

$$\frac{x+3}{1} \div \frac{9-x^2}{x} = \frac{x+3}{1} \cdot \frac{x}{9-x^2}$$

$$= \frac{\cancel{x+3}}{1} \cdot \frac{x}{\cancel{(3+x)}(3-x)}$$

$$= \frac{x}{3-x}$$

6. Simplify by Method 1: Combining fractions in both numerator and denominator.

$$\frac{a+10}{\dfrac{100}{a}-a}$$

151

Copyright © 2013 Pearson Education, Inc.

Example	Student Practice

7. Simplify by Method 2: Multiplying each term of the numerator and denominator by the LCD of all individual fractions.

$$\frac{\dfrac{3}{x+2}+\dfrac{1}{x}}{\dfrac{3}{y}-\dfrac{2}{x}}$$

The LCD of the numerator is $x(x+2)$.
The LCD of the denominator is xy.
Thus, the LCD of the complex fraction is $xy(x+2)$.

$$\frac{\dfrac{3}{x+2}+\dfrac{1}{x}}{\dfrac{3}{y}-\dfrac{2}{x}}\cdot\frac{xy(x+2)}{xy(x+2)}$$

$$=\frac{3xy+xy+2y}{3x(x+2)-2y(x+2)}$$

$$=\frac{4xy+2y}{(x+2)(3x-2y)}=\frac{2y(2x+1)}{(x+2)(3x-2y)}$$

8. Simplify by Method 2: Multiplying each term of the numerator and denominator by the LCD of all individual fractions.

$$\frac{\dfrac{5}{2y+5}+\dfrac{5}{y}}{\dfrac{9}{x}-\dfrac{7}{y}}$$

Extra Practice

1. Simplify by any method. $\dfrac{\dfrac{x^8}{4y^7}}{\dfrac{x^2}{y^3}}$

2. Simplify by any method. $\dfrac{\dfrac{2}{x+3}+1}{\dfrac{1}{3}-\dfrac{2}{x+3}}$

3. Simplify by any method. $\dfrac{\dfrac{4}{a+b}+\dfrac{2}{3}}{\dfrac{b}{a+b}-1}$

4. Simplify by any method. $\dfrac{\dfrac{1}{y+5}-\dfrac{2}{y}}{\dfrac{1}{y}-\dfrac{1}{y+5}}$

Concept Check

Explain how you would simplify the following. $\dfrac{\dfrac{1}{x}+\dfrac{2}{y}}{\dfrac{3}{x}-\dfrac{4}{y}}$

Copyright © 2013 Pearson Education, Inc.

Chapter 6 Rational Expressions and Equations
6.4 Rational Equations

Vocabulary
rational equation   •   extraneous solution   •   no solution   •   apparent solution

1. A(n) _____ is a value received when solving an equation that may or may not be a solution to the equation.

2. An equation has _____ when in the process of solving it, you obtain a contradiction or a value of the variable that does not satisfy the original equation.

3. A(n) _____ is a value received when solving an equation that is not a solution to the equation.

4. A(n) _____ is an equation that has one or more rational expression as terms.

Example	**Student Practice**
1. Solve and check your solution. $$\frac{9}{4} - \frac{1}{2x} = \frac{4}{x}$$	**2.** Solve and check your solution. $$\frac{29}{9} - \frac{2}{3x} = \frac{9}{x}$$

First, multiply each side of the equation by the LCD, which is $4x$. Then, solve.

$$4x\left(\frac{9}{4} - \frac{1}{2x}\right) = 4x\left(\frac{4}{x}\right)$$

$$\cancel{4}x\left(\frac{9}{\cancel{4}}\right) - \cancel{4}\overset{2}{x}\left(\frac{1}{\cancel{2x}}\right) = 4\cancel{x}\left(\frac{4}{\cancel{x}}\right)$$

$$9x - 2 = 16$$

$$9x = 18$$

$$x = 2$$

Check the solution.

$$\frac{9}{4} - \frac{1}{2(2)} \overset{?}{=} \frac{4}{2}$$

$$\frac{9}{4} - \frac{1}{4} \overset{?}{=} 2$$

$$2 = 2$$

Vocabulary Answers: 1. apparent solution 2. no solution 3. extraneous solution 4. rational equation

Copyright © 2013 Pearson Education, Inc.

Example	Student Practice
3. Solve and check. $\dfrac{2}{3x+6} = \dfrac{1}{6} - \dfrac{1}{2x+4}$	**4.** Solve and check. $-\dfrac{5}{2x-6} = \dfrac{1}{8} - \dfrac{11}{4x-12}$

3. (continued)

Factor each denominator and multiply by the LCD, $6(x+2)$. Then, simplify and solve.

$$\frac{2}{3(x+2)} = \frac{1}{6} - \frac{1}{2(x+2)}$$

$$4 = x+2-3$$

$$4 = x-1$$

$$5 = x$$

The check is left to the student.

5. Solve. $\dfrac{y^2-10}{y^2-y-20} = 1 + \dfrac{7}{y-5}$	**6.** Solve. $\dfrac{z^2-14}{z^2-4z-21} = 1 + \dfrac{9}{z-7}$

5. (continued)

Factor each denominator.

$$\frac{y^2-10}{(y-5)(y+4)} = 1 + \frac{7}{y-5}$$

Multiply by the LCD, $(y-5)(y+4)$. Then, simplify and solve.

$$y^2-10 = (y-5)(y+4) + 7(y+4)$$

$$y^2-10 = y^2-y-20+7y+28$$

$$y^2-10 = y^2+6y+8$$

$$-10 = 6y+8$$

$$-18 = 6y$$

$$-3 = y$$

The check is left to the student.

Copyright © 2013 Pearson Education, Inc.

Example	Student Practice
7. Solve. $\dfrac{z+1}{z^2-3z+2}+\dfrac{3}{z-1}=\dfrac{4}{z-2}$	**8.** Solve. $\dfrac{8y+7}{y^2-8y+15}-\dfrac{6}{y-3}=\dfrac{2}{y-5}$

Multiply each term by the LCD, $(z-2)(z-1)$. Then, simplify and solve.

$$\frac{z+1}{z^2-3z+2}+\frac{3}{z-1}=\frac{4}{z-2}$$

$$z+1+3(z-2)=4(z-1)$$

$$z+1+3z-6=4z-4$$

$$4z-5=4z-4$$

$$4z-4z=-4+5$$

$$0=1$$

Of course, $0\neq 1$. Therefore, no value of z makes the original equation true. Hence, the equation has no solution.

| **9.** Solve. $\dfrac{4y}{y+3}-\dfrac{12}{y-3}=\dfrac{4y^2+36}{y^2-9}$ | **10.** Solve. $\dfrac{7x}{x-5}-\dfrac{16}{x+5}=\dfrac{7x^2-15}{x^2-25}$ |

Multiply each term by the LCD, $(y+3)(y-3)$. Then, simplify and solve.

$$\frac{4y}{y+3}-\frac{12}{y-3}=\frac{4y^2+36}{y^2-9}$$

$$4y(y-3)-12(y+3)=4y^2+36$$

$$-24y-36=36$$

$$-24y=72$$

$$y=-3$$

Check the solution.

$$\frac{4(-3)}{-3+3}-\frac{12}{-3-3}\overset{?}{=}\frac{4(-3)^2+36}{(-3)^2-9}$$

$$\frac{-12}{0}-\frac{12}{-6}\overset{?}{=}\frac{36+36}{0}$$

Division by zero is not defined. Thus, this equation has no solution.

Copyright © 2013 Pearson Education, Inc.

Extra Practice

1. Solve. $1 - \dfrac{3}{2x} = \dfrac{7}{4}$

2. Solve. $\dfrac{3}{x-1} + \dfrac{2}{5} = \dfrac{x+15}{5x-5}$

3. Solve. $\dfrac{3}{x^2 - x} = \dfrac{15}{x^2 - 1}$

4. Solve. $\dfrac{30}{y^2 - 25} = \dfrac{6}{y+5} + \dfrac{3}{y-5}$

Concept Check

Why does the equation $\dfrac{1}{x-4} - \dfrac{2}{2x-8} = \dfrac{3}{2}$ have no solution? Explain how this can be determined.

Copyright © 2013 Pearson Education, Inc.

Chapter 6 Rational Expressions and Equations
6.5 Applications: Formulas and Advanced Ratio Exercises

Vocabulary
proportion   •   similar triangles

1. _____ have corresponding angles that are equal and corresponding sides that are proportional.

2. A _____ is an equation that says that two ratios are equal.

Example	Student Practice
1. Solve for a. $\dfrac{1}{f} = \dfrac{1}{a} + \dfrac{1}{b}$	**2.** Solve for b. $\dfrac{1}{f} = \dfrac{1}{a} + \dfrac{1}{b}$

This formula is used in optics in the study of light passing through a lens. It relates the focal length f of the lens to the distance a of an object from the lens and the distance b of the image from the lens.

First multiply each term by the LCD abf. Then simplify. Collect all the terms containing a on one side of the equation. Then factor and solve for a.

$$\frac{1}{f} = \frac{1}{a} + \frac{1}{b}$$

$$ab\!\!\not f\left[\frac{1}{\not f}\right] = \not a bf\left[\frac{1}{\not a}\right] + a\not b f\left[\frac{1}{\not b}\right]$$

$$ab = bf + af$$

$$ab - af = bf$$

$$a(b - f) = bf$$

$$a = \frac{bf}{b - f}$$

Vocabulary Answers: 1. similar triangles 2. proportion

Copyright © 2013 Pearson Education, Inc.

Example	Student Practice
3. The gravitational force F between two masses, m_1 and m_2, a distance d apart is represented by the following formula. Solve for m_2. $F = \dfrac{Gm_1m_2}{d^2}$	**4.** The gravitational force F between two masses, m_1 and m_2, a distance d apart is represented by the following formula. Solve for G. $$F = \dfrac{Gm_1m_2}{d^2}$$

$$F = \dfrac{Gm_1m_2}{d^2}$$

$$d^2[F] = d^2\left[\dfrac{Gm_1m_2}{d^2}\right]$$

$$\dfrac{d^2 F}{Gm_1} = \dfrac{Gm_1m_2}{Gm_1}$$

$$\dfrac{d^2 F}{Gm_1} = m_2$$

5. A company plans to employ 910 people with a ratio of two managers for every 11 workers. How many managers should be hired? How many workers?	**6.** A company plans to employ 1000 people with a ratio of three managers for every 17 workers. How many managers should be hired? How many workers?

If we let $x =$ the number of managers, then $910 - x =$ the number of workers. We are given the ratio of managers to workers, so lets set up our proportion. Two managers over 11 workers equals the number of managers over the number of works.

$$\dfrac{2}{11} = \dfrac{x}{910 - x}$$

Now solve for x.

$$11(910-x)\left[\dfrac{2}{11}\right] = 11(910-x)\left[\dfrac{x}{910-x}\right]$$

$$1820 - 2x = 11x$$

$$140 = x$$

The number of managers needed is 140. The number of workers needed is $910 - x = 910 - 140 = 770$.

Copyright © 2013 Pearson Education, Inc.

Example	Student Practice

7. A helicopter is hovering an unknown distance above an 850-foot building. A man watching the helicopter is 500 feet from the base of the building and 11 feet from a flagpole that is 29 feet tall. The man's line of sight to the helicopter is directly above the flagpole, as you can see in the sketch. How far above the building is the helicopter? Round your answer to the nearest foot.

Note the two triangles in the diagram. We want to find the distance x. The angles at the bases of the triangles are equal. It follows that the top angles must also be equal. Since the angles are equal,

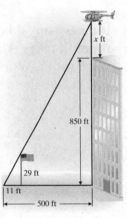

the triangles are similar. Set up the proportion the base of the small triangle over the altitude of the small triangle equals the base of the large triangle over the altitude of the large triangle.

$$\frac{11}{29} = \frac{500}{850 + x}$$

Now solve the equation for x.

$$\frac{11}{29} = \frac{500}{850 + x}$$

$$11(850 + x) = 29(500)$$

$$9350 + 11x = 14,500$$

$$11x = 5150$$

$$x = \frac{5150}{11} \approx 468.18$$

The helicopter is about 468 feet above the building.

8. A helicopter is hovering an unknown distance above an 800-foot building. A man watching the helicopter is 600 feet from the base of the building and 13 feet from a flagpole that is 25 feet tall. The man's line of sight to the helicopter is directly above the flagpole, as you can see in the sketch. How far above the building is the helicopter? Round your answer to the nearest foot.

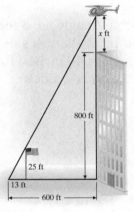

Copyright © 2013 Pearson Education, Inc.

Extra Practice

1. Solve for r. $P = \dfrac{A}{1 + rt}$

2. Solve for g. $\dfrac{s - s_0}{v_0 + gt} = t$

3. The sides of a triangle are 8 feet, 11 feet, and 14 feet. If the shortest side of a similar triangle is 40 inches, find the lengths of the other 2 sides.

4. It takes Amy 4 hours to paint an entire room. It takes Alicia 6 hours to paint the same room. Working together, how long will it take them to paint the room?

Concept Check

Explain how you would solve for H in the following equation.

$S = \dfrac{2A + 3F}{B - H}$

Copyright © 2013 Pearson Education, Inc.

MATH COACH

Mastering the skills you need to do well on the test.

Watch the **MATH COACH** videos in MyMathLab®or on You Tube™ while you work the problems below. These helpful hints will help you avoid making common errors on test problems.

Adding or Subtracting Two or More Rational Expressions—

Problem 6   Simplify. $\dfrac{2}{x^2+5x+6}+\dfrac{3x}{x^2+6x+9}$

> **Helpful Hint:** First, factor each denominator. Form the LCD by taking the product of all the different factors you have obtained. Then multiply the numerator and denominator of each fraction by any factors that do not appear in the denominator.

Did you factor the first denominator into $(x+2)(x+3)$?
Yes ____ No ____
Did you factor the second denominator into $(x+3)(x+3)$?
Yes ____ No ____

If you answered No to either question, stop and carefully factor each denominator.

Did you determine that the LCD is $(x+2)(x+3)(x+3)$?
Yes ____ No ____

Did you multiply the first fraction by $\dfrac{(x+3)}{(x+3)}$?

Yes ____ No ____

Did you multiply the second fraction by $\dfrac{(x+2)}{(x+2)}$? Yes ____ No ____

If you answered No to any of these questions, review how to find the LCD and then multiply each fraction by the correct expression.

If you answered Problem 6 incorrectly, go back and rework the problem using these suggestions.

Simplifying a Complex Rational Expression—Problem 8

Simplify. $\dfrac{\dfrac{1}{x}-\dfrac{3}{x+2}}{\dfrac{2}{x^2+2x}}$

> **Helpful Hint:** You may use either Method 1 or Method 2. We follow Method 1 in the solution to this problem because many students find Method 1 easier to use. For this problem, the two fractions in the numerator need to be combined, but the denominator does not involve combining fractions.

Did you find the LCD of the two fractions in the numerator to be $x(x+2)$? Yes ____ No ____

Did you multiply the first fraction in the numerator by $\dfrac{(x+2)}{(x+2)}$? Yes ____ No ____

Did you multiply the second fraction in the numerator by $\dfrac{x}{x}$? Yes ____ No ____

If you answered No to any of these questions, stop and review how to find the LCD. Consider that when combining two fractions, we multiply the numerator and denominator of each fraction by any factor in the LCD that is not presently in

the denominator. Please go back and complete these steps again.

Did you combine the two fractions in the numerator to obtain $\dfrac{-2x+2}{x(x+2)}$?

Yes ____ No ____

If you answered No, remember that you need to combine $x+2-3x$ to obtain $-2x+2$. Now complete the division process just as you would divide two fractions.

Now go back and rework the problem using these suggestions.

Copyright © 2013 Pearson Education, Inc.

Solving a Rational Equation—Problem 11   Solve for the variable and check your answers.

Helpful Hint: Factor the denominators. Find the LCD. Multiply each fraction on both sides of the equation by the LCD. The resulting equation should not contain any fractions. Simplify and solve for the variable.

$$\frac{1}{2y+4}-\frac{1}{6}=\frac{-2}{3y+6}$$

Did you find that the LCD is $(2)(3)(y+2)$ or $6(y+2)$?
Yes ____   No ____

If you answered No, remember that the denominator of the first fraction factors to $2(y+2)$, the denominator of the second fraction factors to $2\cdot3$, and the denominator of the third fraction factors to $3(y+2)$. Work the problem again to find the correct LCD.

When you multiplied the LCD by each fraction, did you obtain the unsimplified equation $3-(y+2)=2(-2)$?
Yes ____   No ____

If you answered No, carefully multiply $6(y+2)$ by each of the three fractions. Make sure that your resulting equation is cleared of all fractions.

Then simplify the equation and solve for y. Check your solution for y by substituting the value in the original equation.

If you answered Problem 11 incorrectly, go back and rework the problem using these suggestions.

Solving an Applied Problem Involving Ratios—Problem 15   A total of 286 employees at Kaiser Telecommunication Systems were eligible this year for a high-performance bonus. The company president announced that for every three employees who got the bonus, nineteen employees did not. If the president was correct, how many employees got the high-performance bonus? How many did not get the high-performance bonus?

Helpful Hint: Read through the problem carefully. Make sure you can identify a variable expression for the number of employees who got a bonus and the number of employees who did not get a bonus. Setting up an equation requires that you first find these two expressions.

If you let x be the number of employees who got a bonus, can you determine that $286-x$ is the number of employees who did not get a bonus? Yes ____   No ____

If you answered No, remember that out of a total of 286 employees, some received bonuses and the remaining number did not. So if x people got a bonus, then $286-x$ did not get a bonus.

Did you set up a rational equation with one fraction as the ratio of $\frac{3}{19}$ and the other fraction as $\frac{x}{286-x}$?
Yes ____   No ____

If you answered No, remember that if you choose x as one of the numerators, then 3 and x should be the two numerators. This means that 19 and $286-x$ should be the two denominators, respectively. Another possibility is to set up the equation $\frac{19}{3}=\frac{286-x}{x}$.

The key is to have the same value represented for the numerator and the same value represented for the denominator of each fraction.

After multiplying by the LCD, did you obtain the equation $3(286-x)=19x$ or $19x=3(286-x)$? Yes ____   No ____

If you answered No, remember that you must set the two ratios equal to each other and then multiply by the LCD to get an equation without fractions. Now solve the equation for x. Remember to find the values for both x and $286-x$ in your final answer.

If you answered Problem 15 incorrectly, go back and rework the problem using these suggestions.

Copyright © 2013 Pearson Education, Inc.

Chapter 7 Rational Exponents
7.1 Simplifying Expressions with Rational Exponents

Vocabulary
rational exponents   •   denominator   •   quotient rule   •   exponential factor

1. You need to change fractional exponents to equivalent fractional exponents with the same _____ when the rules of exponents require you to add or subtract them.

2. Exponents that are fractions are called _____.

3. When factoring an expression, if the terms contain exponents, we look for the same _____ in each term.

Example	Student Practice
1. Simplify. $\left(\dfrac{5xy^{-3}}{2x^{-4}y}\right)^{-2}$	**2.** Simplify. $\left(\dfrac{4a^{-3}b}{3a^{-2}b^{-5}}\right)^{-3}$

Apply the rules of exponents.

$$\left(\frac{5xy^{-3}}{2x^{-4}y}\right)^{-2} = \frac{\left(5xy^{-3}\right)^{-2}}{\left(2x^{-4}y\right)^{-2}} \qquad \left(\frac{x}{y}\right)^{n} = \frac{x^{n}}{y^{n}}$$

$$= \frac{5^{-2}x^{-2}\left(y^{-3}\right)^{-2}}{2^{-2}\left(x^{-4}\right)^{-2}y^{-2}} \qquad (xy)^{n} = x^{n}y^{n}$$

$$= \frac{5^{-2}x^{-2}y^{6}}{2^{-2}x^{8}y^{-2}} \qquad \left(x^{m}\right)^{n} = x^{mn}$$

$$= \frac{5^{-2}}{2^{-2}} \cdot \frac{x^{-2}}{x^{8}} \cdot \frac{y^{6}}{y^{-2}}$$

$$= \frac{2^{2}}{5^{2}} \cdot \frac{x^{-2}}{x^{8}} \cdot \frac{y^{6}}{y^{-2}} \qquad \frac{x^{-n}}{y^{-m}} = \frac{y^{m}}{x^{n}}$$

$$= \frac{2^{2}}{5^{2}} \cdot x^{-2-8} \cdot y^{6+2} \qquad \frac{x^{m}}{x^{n}} = x^{m-n}$$

$$= \frac{4}{25}x^{-10}y^{8}$$

Vocabulary Answers: 1. denominator 2. rational exponents 3. exponential factor

Copyright © 2013 Pearson Education, Inc.

Example	Student Practice
3. Simplify.	**4.** Simplify.
(a) $\left(x^{2/3}\right)^4$	**(a)** $\left(x^3\right)^{4/5}$
$\left(x^{2/3}\right)^4 = x^{(2/3)(4/1)} = x^{8/3}$	
(b) $\dfrac{x^{5/6}}{x^{1/6}}$	**(b)** $\dfrac{x^{7/8}}{x^{1/8}}$
$\dfrac{x^{5/6}}{x^{1/6}} = x^{5/6-1/6} = x^{4/6} = x^{2/3}$	**(c)** $6^{5/13} \cdot 6^{4/13}$
(c) $x^{2/3} \cdot x^{-1/3}$	
$x^{2/3} \cdot x^{-1/3} = x^{2/3-1/3} = x^{1/3}$	

Example	Student Practice
5. Simplify. Express your answers with positive exponents only.	**6.** Simplify. Express your answers with positive exponents only.
(a) $\left(2x^{1/2}\right)\left(3x^{1/3}\right)$	**(a)** $\left(-3x^{2/5}\right)\left(4x^{1/2}\right)$
$\left(2x^{1/2}\right)\left(3x^{1/3}\right) = 6x^{1/2+1/3}$	
$= 6x^{3/6+2/6} = 6x^{5/6}$	
(b) $\dfrac{18x^{1/4}y^{-1/3}}{-6x^{-1/2}y^{1/6}}$	**(b)** $\dfrac{25x^{3/4}y^{-5/12}}{5x^{-1/8}y^{1/4}}$
$\dfrac{18x^{1/4}y^{-1/3}}{-6x^{-1/2}y^{1/6}} = -3x^{1/4-(-1/2)}y^{-1/3-1/6}$	
$= -3x^{1/4+2/4}y^{-2/6-1/6}$	
$= -3x^{3/4}y^{-3/6}$	
$= -3x^{3/4}y^{-1/2}$	
$= -\dfrac{3x^{3/4}}{y^{1/2}}$	

Copyright © 2013 Pearson Education, Inc.

Example	Student Practice
7. Multiply and simplify. $$-2x^{5/6}\left(3x^{1/2} - 4x^{-1/3}\right)$$ We will need to be very careful when we add the exponents for x as we use the distributive property. $$-2x^{5/6}\left(3x^{1/2} - 4x^{-1/3}\right)$$ $$= -6x^{5/6+1/2} + 8x^{5/6-1/3}$$ $$= -6x^{5/6+3/6} + 8x^{5/6-2/6}$$ $$= -6x^{8/6} + 8x^{3/6}$$ $$= -6x^{4/3} + 8x^{1/2}$$	**8.** Multiply and simplify. $$4x^{2/3}\left(-2x^{1/2} + 6x^{-1/6}\right)$$
9. Evaluate. **(a)** $(25)^{3/2}$ $$(25)^{3/2} = \left(5^2\right)^{3/2} = 5^{2/1\cdot3/2} = 5^3 = 125$$ **(b)** $(27)^{2/3}$ $$(27)^{2/3} = \left(3^3\right)^{2/3} = 3^{3/1\cdot2/3} = 3^2 = 9$$	**10.** Evaluate. **(a)** $(16)^{3/2}$ **(b)** $(64)^{2/3}$
11. Write as one fraction with positive exponents. $2x^{-1/2} + x^{1/2}$ $$2x^{-1/2} + x^{1/2} = \frac{2}{x^{1/2}} + \frac{x^{1/2} \cdot x^{1/2}}{x^{1/2}}$$ $$= \frac{2}{x^{1/2}} + \frac{x^1}{x^{1/2}}$$ $$= \frac{2+x}{x^{1/2}}$$	**12.** Write as one fraction with positive exponents. $5x^{-3/4} + x^{1/4}$

Copyright © 2013 Pearson Education, Inc.

Example	Student Practice
13. Factor out the common factor $2x$. $$2x^{3/2} + 4x^{5/2}$$ Rewrite the exponent of each term so that each term contains the factor $2x$ or $2x^{2/2}$. $$2x^{3/2} + 4x^{5/2}$$ $$= 2x^{2/2+1/2} + 4x^{2/2+3/2}$$ $$= 2\left(x^{2/2}\right)\left(x^{1/2}\right) + 4\left(x^{2/2}\right)\left(x^{3/2}\right)$$ $$= 2x\left(x^{1/2} + 2x^{3/2}\right)$$	**14.** Factor out the common factor $6z$. $$18z^{5/3} - 24z^{4/3}$$

Extra Practice

1. Simplify. Express your answers with positive exponents only.

$$\left(x^{4/5}\right)^4$$

2. Simplify. Express your answers with positive exponents only.

$$\left(a^{5/4}b^{2/3}\right)\left(a^{-1/4}b^{2/3}\right)$$

3. Write as one fraction with positive exponents.

$$y^{-2/3} + 3y^{1/3}$$

4. Factor out the common factor $5x$.

$$10x^{7/4} + 25x^{9/8}$$

Concept Check

Explain how you would simplify the following. $x^{9/10} \cdot x^{-1/5}$

Copyright © 2013 Pearson Education, Inc.

Chapter 7 Rational Exponents
7.2 Radical Expressions and Functions

Vocabulary
square root • principal square root • radical sign • radical expression • radicand
index • higher-order roots • cube root • fourth root • square root function

1. Because the symbol $\sqrt{x}$ represents exactly one real number for all real numbers x that are nonnegative, we can use it to define the _____, $f(x) = \sqrt{x}$.

2. The positive square root is the _____.

3. A(n) _____ of a number is a value that when cubed is equal to the original number.

4. In the radical expression $\sqrt[n]{x}$, x is the radicand and n is the _____.

Example	Student Practice
1. Find the indicated function values of the function $f(x) = \sqrt{2x+4}$. Round your answers to the nearest tenth when necessary.	**2.** Find the indicated function values of the function $f(x) = \sqrt{5x-4}$. Round your answers to the nearest tenth when necessary.
(a) $f(-2)$ Substitute -2 for x in the function. $f(-2) = \sqrt{2(-2)+4} = \sqrt{-4+4}$ $\quad\quad = \sqrt{0} = 0$	**(a)** $f(4)$
(b) $f(6)$ $f(6) = \sqrt{2(6)+4} = \sqrt{12+4}$ $\quad\quad = \sqrt{16} = 4$	**(b)** $f(1)$
(c) $f(3)$ $f(3) = \sqrt{2(3)+4} = \sqrt{6+4}$ $\quad\quad = \sqrt{10} \approx 3.2$	**(c)** $f(3)$

Vocabulary Answers: 1. square root function 2. principal square root 3. cube root 4. index

Copyright © 2013 Pearson Education, Inc.

Example	Student Practice
3. Find the domain of the function. $f(x) = \sqrt{3x - 6}$ The expression $3x - 6$ must be nonnegative. $3x - 6 \geq 0$ $\quad 3x \geq 6$ $\quad\;\; x \geq 2$ Thus, the domain is all real numbers x, where $x \geq 2$.	**4.** Find the domain of the function. $f(x) = \sqrt{\dfrac{2}{3}x - 6}$
5. Change $\sqrt[4]{x^4}$ to rational exponents and simplify. Assume that all variables are nonnegative real numbers. If n is a positive integer and x is a nonnegative real number, then $x^{1/n} = \sqrt[n]{x}$. $\sqrt[4]{x^4} = \left(x^4\right)^{1/4} = x^{4/4} = x^1 = x$	**6.** Change $\sqrt[6]{7^6}$ to rational exponents and simplify. Assume that all variables are nonnegative real numbers.
7. Replace all radicals with rational exponents. **(a)** $\sqrt[3]{x^2}$ $\quad \sqrt[3]{x^2} = \left(x^2\right)^{1/3}$ $\qquad\;\;\; = x^{2/3}$ **(b)** $\left(\sqrt[5]{w}\right)^7$ $\quad \left(\sqrt[5]{w}\right)^7 = \left(w^{1/5}\right)^7$ $\qquad\qquad\;\; = w^{7/5}$	**8.** Replace all radicals with rational exponents. **(a)** $\sqrt[5]{z^3}$ **(b)** $\left(\sqrt[3]{a}\right)^8$

Copyright © 2013 Pearson Education, Inc.

Example	Student Practice
9. Change to radical form.	**10.** Change to radical form.
(a) $w^{-2/3}$	**(a)** $\left(ab\right)^{-5/3}$

For positive integers m and n and any real number x where $x \neq 0$ and for which $x^{1/n}$ is defined, then

$$x^{-m/n} = \frac{1}{x^{m/n}} = \frac{1}{\left(\sqrt[n]{x}\right)^m} = \frac{1}{\sqrt[n]{x^m}}.$$

$$w^{-2/3} = \frac{1}{w^{2/3}} = \frac{1}{\left(\sqrt[3]{w}\right)^2} = \frac{1}{\sqrt[3]{w^2}}$$

(b) $\left(3x\right)^{3/4}$ **(b)** $\left(5x\right)^{1/5}$

For positive integers m and n and any real number x for which $x^{1/n}$ is defined, then $x^{m/n} = \left(\sqrt[n]{x}\right)^m = \sqrt[n]{x^m}$.

$$\left(3x\right)^{3/4} = \sqrt[4]{\left(3x\right)^3} = \sqrt[4]{27x^3} \text{ or}$$

$$\left(3x\right)^{3/4} = \left(\sqrt[4]{3x}\right)^3$$

Example	Student Practice
11. Change to radical form and evaluate.	**12.** Change to radical form and evaluate.
(a) $\left(-16\right)^{5/2}$	**(a)** $64^{-1/3}$

$\left(-16\right)^{5/2} = \left(\sqrt{-16}\right)^5$; however, $\sqrt{-16}$ is not a real number. Thus, $\left(-16\right)^{5/2}$ is not a real number.

(b) $\left(-81\right)^{1/4}$

(b) $144^{-1/2}$

$$144^{-1/2} = \frac{1}{144^{1/2}} = \frac{1}{\sqrt{144}} = \frac{1}{12}$$

Copyright © 2013 Pearson Education, Inc.

Example	Student Practice
13. Simplify. Assume that x and y may be any real numbers.	**14.** Simplify. Assume that x and y may be any real numbers.

(a) $\sqrt{49x^2}$

Since the index is even, take the absolute value.

$$\sqrt{49x^2} = 7|x|$$

(a) $\sqrt{25a^2}$

(b) $\sqrt[4]{81y^{16}}$

$$\sqrt[4]{81y^{16}} = 3|y^4|$$

Since any number raised to the fourth is positive we can write it as $\sqrt[4]{81y^{16}} = 3y^4$.

(b) $\sqrt[3]{64z^6}$

(c) $\sqrt[4]{81x^8y^4}$

(c) $\sqrt[3]{27x^6y^9}$

Note that the index is odd, thus we do not need the absolute value.

$$\sqrt[3]{27x^6y^9} = \sqrt[3]{3^3\left(x^2\right)^3\left(y^3\right)^3} = 3x^2y^3$$

Extra Practice

1. Evaluate if possible. $\sqrt[15]{(7)^{15}}$

2. Replace all radicals with rational exponents. Assume that variables represent positive real numbers.

$$\sqrt[8]{(4m-3n)^5}$$

3. Change to radical form. $4^{-4/7}$

4. Simplify. $\sqrt{49a^6b^{16}}$

Concept Check

Explain how you would simplify the following. $\sqrt[4]{81x^8y^{16}}$

Copyright © 2013 Pearson Education, Inc.

Name: _____   Date: _____

Instructor: _____   Section: _____

Chapter 7 Rational Exponents
7.3 Simplifying, Adding, and Subtracting Radicals

Vocabulary
product rule   •   radicand   •   index   •   like radicals   •   distributive property

1. Like radicals have the same _____ and index.

2. The_____ for radicals states that for all nonnegative real numbers a and b and positive integers n, $\sqrt[n]{a}\sqrt[n]{b} = \sqrt[n]{ab}$.

3. Only _____ can be added or subtracted.

Example	Student Practice
1. Simplify. $\sqrt{32}$ Recall that for all nonnegative real numbers a and b and positive integers n, $\sqrt[n]{a}\sqrt[n]{b} = \sqrt[n]{ab}$. Factor the radicand into the largest perfect square factors possible, then simplify. $\sqrt{32} = \sqrt{16 \cdot 2} = \sqrt{16}\sqrt{2} = 4\sqrt{2}$	**2.** Simplify. $\sqrt{50}$
3. Simplify. $\sqrt{48}$ $\sqrt{48} = \sqrt{16 \cdot 3} = \sqrt{16}\sqrt{3} = 4\sqrt{3}$	**4.** Simplify. $\sqrt{108}$
5. Simplify. $\sqrt[3]{-81}$ Factor the radicand into the largest perfect cube factors possible, then simplify. $\sqrt[3]{-81} = \sqrt[3]{-27}\sqrt[3]{3} = -3\sqrt[3]{3}$	**6.** Simplify. $\sqrt[3]{-162}$

Vocabulary Answers: 1. radicand 2. product rule 3. like radicals

Copyright © 2013 Pearson Education, Inc.

Example	Student Practice
7. Simplify.	**8.** Simplify.

7. Simplify.

(a) $\sqrt{27x^3 y^4}$

Factor out the perfect squares.

$$\sqrt{27x^3 y^4} = \sqrt{9 \cdot 3 \cdot x^2 \cdot x \cdot y^4}$$
$$= \sqrt{9x^2 y^4}\sqrt{3x} = 3xy^2\sqrt{3x}$$

(b) $\sqrt[3]{16x^4 y^3 z^6}$

Factor out the perfect cubes.

$$\sqrt[3]{16x^4 y^3 z^6} = \sqrt[3]{8 \cdot 2 \cdot x^3 \cdot x \cdot y^3 \cdot z^6}$$
$$= \sqrt[3]{8x^3 y^3 z^6}\sqrt[3]{2x}$$
$$= 2xyz^2\sqrt[3]{2x}$$

8. Simplify.

(a) $\sqrt{32a^5 b^4}$

(b) $\sqrt[3]{192x^2 y^4 z^5}$

---

9. Combine. $2\sqrt{5} + 3\sqrt{5} - 4\sqrt{5}$

All three radicals are like radicals because they have the same radicand and index. Thus, they may be combined.

$$2\sqrt{5} + 3\sqrt{5} - 4\sqrt{5} = (2 + 3 - 4)\sqrt{5}$$
$$= 1\sqrt{5} = \sqrt{5}$$

10. Combine. $6\sqrt{3a} - 3\sqrt{3a} + \sqrt{3a}$.

---

11. Combine. $5\sqrt{3} - \sqrt{27} + 2\sqrt{48}$

Sometimes when simplifying radicands, you may find like radicals.

$$5\sqrt{3} - \sqrt{27} + 2\sqrt{48}$$
$$= 5\sqrt{3} - \sqrt{9}\sqrt{3} + 2\sqrt{16}\sqrt{3}$$
$$= 5\sqrt{3} - 3\sqrt{3} + 2(4)\sqrt{3}$$
$$= 5\sqrt{3} - 3\sqrt{3} + 8\sqrt{3}$$
$$= 10\sqrt{3}$$

12. Combine. $6\sqrt{5} + \sqrt{80} - 3\sqrt{20}$

Copyright © 2013 Pearson Education, Inc.

Example	Student Practice
13. Combine. $6\sqrt{x}+4\sqrt{12x}-\sqrt{75x}+3\sqrt{x}$	**14.** Combine. $3\sqrt{a}-2\sqrt{32a}+\sqrt{50a}+2\sqrt{a}$

$6\sqrt{x}+4\sqrt{12x}-\sqrt{75x}+3\sqrt{x}$

$=6\sqrt{x}+4\sqrt{4}\sqrt{3x}-\sqrt{25}\sqrt{3x}+3\sqrt{x}$

$=6\sqrt{x}+8\sqrt{3x}-5\sqrt{3x}+3\sqrt{x}$

$=6\sqrt{x}+3\sqrt{x}+8\sqrt{3x}-5\sqrt{3x}$

$=9\sqrt{x}+3\sqrt{3x}$

---

15. Combine. $2\sqrt[3]{81x^3y^4}+3xy\sqrt[3]{24y}$

$2\sqrt[3]{81x^3y^4}+3xy\sqrt[3]{24y}$

$=2\sqrt[3]{27x^3y^3}\sqrt[3]{3y}+3xy\sqrt[3]{8}\sqrt[3]{3y}$

$=2(3xy)\sqrt[3]{3y}+3xy(2)\sqrt[3]{3y}$

$=6xy\sqrt[3]{3y}+6xy\sqrt[3]{3y}$

$=12xy\sqrt[3]{3y}$

16. Combine. $\sqrt[3]{40x^6z^4}+3x^2\sqrt[3]{320z^4}$

Extra Practice

1. Simplify. Assume that all variables are nonnegative real numbers. $\sqrt{25x^3}$

2. Simplify. Assume that all variables are nonnegative real numbers. $\sqrt[4]{625ab^{19}}$

3. Combine. $6\sqrt{7}-4\sqrt{5}+4\sqrt{7}$

4. Combine. Assume that all variables represent nonnegative real numbers. $6\sqrt{48x^2}-2\sqrt{27x^2}-\sqrt{3x^2}$

Concept Check

Explain how you would simplify the following. $\sqrt[4]{16x^{13}y^{16}}$

Copyright © 2013 Pearson Education, Inc.

Copyright © 2013 Pearson Education, Inc.

Chapter 7 Rational Exponents
7.4 Multiplying and Dividing Radicals

Vocabulary
product rule for radicals   •   FOIL method   •   quotient rule for radicals
rationalizing the denominator   •   conjugates

1. The expressions $a+b$ and $a-b$, where a and b represent any algebraic term, are called _____.

2. The _____ states that for all nonnegative real numbers a, all positive numbers b, and positive integers n, $\dfrac{\sqrt[n]{a}}{\sqrt[n]{b}} = \sqrt[n]{\dfrac{a}{b}}$.

3. _____ is the process of transforming a fraction with one or more radicals in the denominator into an equivalent fraction without a radical in the denominator.

Example	Student Practice
1. Multiply. $\left(3\sqrt{2}\right)\left(5\sqrt{11x}\right)$ To multiply, use the product rule for radicals, $\sqrt[n]{a}\sqrt[n]{b} = \sqrt[n]{ab}$. $\left(3\sqrt{2}\right)\left(5\sqrt{11x}\right) = (3)(5)\left(\sqrt{2\cdot 11x}\right)$ $\qquad = 15\sqrt{2\cdot 11x} = 15\sqrt{22x}$	**2.** Multiply. $\left(3\sqrt{7}\right)\left(-4\sqrt{3z}\right)$
3. Multiply. $\left(\sqrt{2}+3\sqrt{5}\right)\left(2\sqrt{2}-\sqrt{5}\right)$ To multiply two binomials containing radicals, we can use the distributive property or the FOIL method. For this problem, we use the FOIL method. $\left(\sqrt{2}+3\sqrt{5}\right)\left(2\sqrt{2}-\sqrt{5}\right)$ $= 2\sqrt{4}-\sqrt{10}+6\sqrt{10}-3\sqrt{25}$ $= 4+5\sqrt{10}-15 = -11+5\sqrt{10}$	**4.** Multiply. $\left(4\sqrt{3}-3\sqrt{5}\right)\left(2\sqrt{3}-\sqrt{5}\right)$

Vocabulary Answers: 1. conjugates 2. quotient rule for radicals 3. rationalizing the denominator

Copyright © 2013 Pearson Education, Inc.

Example	Student Practice
5. Multiply. $\left(7-3\sqrt{2}\right)\left(4-\sqrt{3}\right)$ Use the FOIL method. $\left(7-3\sqrt{2}\right)\left(4-\sqrt{3}\right)$ $=28-7\sqrt{3}-12\sqrt{2}+3\sqrt{6}$	**6.** Multiply. $\left(6+4\sqrt{3}\right)\left(2+3\sqrt{2}\right)$
7. Multiply. $\left(\sqrt{7}+\sqrt{3x}\right)^2$ Use $\left(a+b\right)^2 = a^2 + 2ab + b^2,$ where $a = \sqrt{7}$ and $b = \sqrt{3x}.$ $\left(\sqrt{7}+\sqrt{3x}\right)^2$ $=\left(\sqrt{7}\right)^2 + 2\sqrt{7}\sqrt{3x} + \left(\sqrt{3x}\right)^2$ $=7 + 2\sqrt{21x} + 3x$	**8.** Multiply. $\left(\sqrt{11}-\sqrt{2a}\right)^2$
9. Divide. **(a)** $\sqrt[3]{\dfrac{125}{8}}$ For all nonnegative real numbers $a,$ all positive numbers $b,$ and positive integers $n,$ $\dfrac{\sqrt[n]{a}}{\sqrt[n]{b}} = \sqrt[n]{\dfrac{a}{b}}.$ $\sqrt[3]{\dfrac{125}{8}} = \dfrac{\sqrt[3]{125}}{\sqrt[3]{8}} = \dfrac{5}{2}$ **(b)** $\dfrac{\sqrt{28x^5y^3}}{\sqrt{7x}}$ $\dfrac{\sqrt{28x^5y^3}}{\sqrt{7x}} = \sqrt{\dfrac{28x^5y^3}{7x}} = \sqrt{4x^4y^3}$ $= 2x^2y\sqrt{y}$	**10.** Divide. **(a)** $\sqrt[3]{\dfrac{8}{27}}$ **(b)** $\dfrac{\sqrt{72a^3b^7}}{\sqrt{2b^3}}$

Copyright © 2013 Pearson Education, Inc.

Example	Student Practice
11. Simplify. $\dfrac{3}{\sqrt{12x}}$	**12.** Simplify. $\dfrac{5}{\sqrt{32z}}$

Simplify the radical in the denominator, then multiply in order to rationalize the denominator.

$$\frac{3}{\sqrt{12x}} = \frac{3}{\sqrt{4}\sqrt{3x}}$$

$$= \frac{3}{2\sqrt{3x}} \cdot \frac{\sqrt{3x}}{\sqrt{3x}}$$

$$= \frac{\cancel{3}\sqrt{3x}}{2(\cancel{3}x)} = \frac{\sqrt{3x}}{2x}$$

---

13. Simplify. $\sqrt[3]{\dfrac{2}{3x^2}}$

14. Simplify. $\sqrt[3]{\dfrac{54}{2a}}$

Multiply the numerator and denominator by a value that will make the radicand in the denominator a perfect cube (i.e., rationalize the denominator). In this case, multiply the numerator and denominator by $9x$ because $9x \cdot 3x^2 = 27x^3$ which is a perfect cube.

$$\sqrt[3]{\frac{2}{3x^2}} = \sqrt[3]{\frac{2}{3x^2} \cdot \frac{9x}{9x}}$$

$$= \sqrt[3]{\frac{18x}{27x^3}}$$

Rewrite with the quotient rule and simplify.

$$\sqrt[3]{\frac{18x}{27x^3}} = \frac{\sqrt[3]{18x}}{\sqrt[3]{27x^3}}$$

$$= \frac{\sqrt[3]{18x}}{3x}$$

Copyright © 2013 Pearson Education, Inc.

Example	Student Practice
15. Simplify. $\dfrac{5}{3+\sqrt{2}}$	**16.** Simplify. $\dfrac{\sqrt{5}+3\sqrt{7}}{2\sqrt{5}-\sqrt{7}}$

Multiply the numerator and denominator by the conjugate of $3+\sqrt{2}$.

$$\frac{5}{3+\sqrt{2}} = \frac{5}{3+\sqrt{2}} \cdot \frac{3-\sqrt{2}}{3-\sqrt{2}}$$

$$= \frac{15-5\sqrt{2}}{3^2-\left(\sqrt{2}\right)^2}$$

$$= \frac{15-5\sqrt{2}}{9-2} = \frac{15-5\sqrt{2}}{7}$$

Extra Practice

1. Multiply and simplify. Assume that all variables are nonnegative numbers.

$$\left(3a\sqrt{a}\right)\left(5\sqrt{b}\right)$$

2. Multiply and simplify. Assume that all variables are nonnegative numbers.

$$\left(5\sqrt{7}+3\sqrt{10}\right)^2$$

3. Divide and simplify. Assume that all variables represent positive numbers.

$$\sqrt{\frac{18x}{25y^4}}$$

4. Simplify by rationalizing the denominator. $\dfrac{2x}{\sqrt{5}-\sqrt{3}}$

Concept Check

Explain how you would rationalize the denominator of the following. $\dfrac{5\sqrt{3}-3\sqrt{2}}{3\sqrt{2}-2\sqrt{3}}$

Copyright © 2013 Pearson Education, Inc.

Chapter 7 Rational Exponents
7.5 Radical Equations

Vocabulary

radical equation • isolating the radical term • zero factor • extraneous solution

1. If the result of squaring both sides of a radical equation is a quadratic equation, collect all terms on one side and use the _____ method to solve.

2. A(n) _____ is an equation with a variable in one or more of the radicals.

3. Getting one radical expression alone on one side of the equation is referred to as _____.

Example	Student Practice
1. Solve. $\sqrt{2x+9} = x+3$	**2.** Solve. $\sqrt{3x+7} = x-1$

Example

1. Solve. $\sqrt{2x+9} = x+3$

Square each side and simplify.
$$\left(\sqrt{2x+9}\right)^2 = (x+3)^2$$
$$2x+9 = x^2+6x+9$$

Collect all terms on one side and factor.
$$2x+9 = x^2+6x+9$$
$$0 = x^2+4x$$
$$0 = x(x+4)$$

Set each factor equal to zero and solve.
$$x=0 \quad \text{or} \quad x+4=0$$
$$x=0 \quad\quad\quad\quad x=-4$$

Check.
$$x=0: \quad \sqrt{2(0)+9} \overset{?}{=} (0)+3$$
$$3=3$$
$$x=-4: \quad \sqrt{2(-4)+9} \overset{?}{=} (-4)+3$$
$$1 \neq -1$$

Thus, 0 is the only solution.

Vocabulary Answers: 1. zero factor 2. radical equation 3. isolating the radical

Copyright © 2013 Pearson Education, Inc.

Example	Student Practice
3. Solve. $\sqrt{10x+5}-1=2x$	**4.** Solve. $3x=-1+\sqrt{30x+10}$

Example

3. Solve. $\sqrt{10x+5}-1=2x$

Isolate the radical term. Then square each side and simplify.

$$\sqrt{10x+5}-1=2x$$
$$\sqrt{10x+5}=2x+1$$
$$\left(\sqrt{10x+5}\right)^2=(2x+1)^2$$
$$10x+5=4x^2+4x+1$$

Collect all terms on one side and factor.
$$10x+5=4x^2+4x+1$$
$$0=4x^2-6x-4$$
$$0=2\left(2x^2-3x-2\right)$$
$$0=2(2x+1)(x-2)$$

Set each factor equal to zero and solve.
$$2x+1=0 \qquad\qquad x-2=0$$
$$2x=-1 \qquad\qquad x=2$$
$$x=-\frac{1}{2}$$

Check.

$$x=-\frac{1}{2}:\quad \sqrt{10\left(-\frac{1}{2}\right)+5}-1\overset{?}{=}2\left(-\frac{1}{2}\right)$$
$$\sqrt{-5+5}-1\overset{?}{=}-1$$
$$\sqrt{0}-1\overset{?}{=}-1$$
$$-1=-1$$
$$x=2:\quad \sqrt{10(2)+5}-1\overset{?}{=}2(2)$$
$$\sqrt{25}-1\overset{?}{=}4$$
$$4=4$$

Both answers check. Thus, the solutions are $-\frac{1}{2}$ and 2.

Copyright © 2013 Pearson Education, Inc.

Example	Student Practice
5. Solve. $\sqrt{5x+1}-\sqrt{3x}=1$	**6.** Solve. $\sqrt{x+4}-2=\sqrt{x-4}$

5. Solve. $\sqrt{5x+1}-\sqrt{3x}=1$

Isolate one of the radicals.

$$\sqrt{5x+1}-\sqrt{3x}=1$$
$$\sqrt{5x+1}=1+\sqrt{3x}$$

Square each side.

$$\sqrt{5x+1}=1+\sqrt{3x}$$
$$\left(\sqrt{5x+1}\right)^2=\left(1+\sqrt{3x}\right)^2$$
$$5x+1=\left(1+\sqrt{3x}\right)\left(1+\sqrt{3x}\right)$$
$$5x+1=1+2\sqrt{3x}+3x$$

Isolate the remaining radical.

$$5x+1=1+2\sqrt{3x}+3x$$
$$2x=2\sqrt{3x}$$
$$x=\sqrt{3x}$$

Square each side.

$$x=\sqrt{3x}$$
$$\left(x\right)^2=\left(\sqrt{3x}\right)^2$$
$$x^2=3x$$

Collect all terms on one side and factor.
$$x^2=3x$$
$$x^2-3x=0$$
$$x(x-3)=0$$
$$x=0 \quad\text{or}\quad x-3=0$$
$$x=0 \qquad\qquad x=3$$

The check is left to the student. Verify that both 0 and 3 are valid solutions.

Copyright © 2013 Pearson Education, Inc.

Example	Student Practice
7. Solve. $\sqrt{2y+5}-\sqrt{y-1}=\sqrt{y+2}$	**8.** Solve. $\sqrt{y+6}-\sqrt{y-2}=\sqrt{2y-2}$

$$\sqrt{2y+5}-\sqrt{y-1}=\sqrt{y+2}$$

$$\left(\sqrt{2y+5}-\sqrt{y-1}\right)^2=\left(\sqrt{y+2}\right)^2$$

Expand both sides of the equation.

$$2y+5-2\sqrt{(y-1)(2y+5)}+y-1=y+2$$

Isolate the radical and solve for y.

$$-2\sqrt{(y-1)(2y+5)}=-2y-2$$

$$\sqrt{(y-1)(2y+5)}=y+1$$

$$\left(\sqrt{(y-1)(2y+5)}\right)^2=(y+1)^2$$

$$2y^2+3y-5=y^2+2y+1$$

$$y^2+y-6=0$$

$$(y+3)(y-2)=0$$

$$y=-3 \quad\text{or}\quad y=2$$

The check is left to the student. Verify that 2 is a valid solution but -3 is not.

Extra Practice

1. Solve the radical equation. Check your solution. $\sqrt{10x-5}=5$

2. Solve the radical equation. Check your solution. $\sqrt{x+4}+8=3$

3. Solve the radical equation. This will usually involve squaring each side twice. Check your solution. $\sqrt{3x+4}=2+\sqrt{x}$

4. Solve the radical equation. This will usually involve squaring each side twice. Check your solution.
$$\sqrt{2x+6}-\sqrt{x+4}=\sqrt{x-4}$$

Concept Check

When you try to solve the equation $2+\sqrt{x+10}=x$, you obtain the values $x=-1$ and $x=6$. Explain how you would determine if either of these values is a solution of the radical equation.

Copyright © 2013 Pearson Education, Inc.

Chapter 7 Rational Exponents
7.6 Complex Numbers

Vocabulary
imaginary number • complex number • real part • imaginary part • conjugates

1. The complex numbers $a+bi$ and $a-bi$ are called _____.

2. The _____ i is defined as $i = \sqrt{-1}$ and $i^2 = -1$.

3. A number that can be written in the form $a+bi$, where a and b are real numbers, is a
    _____.

4. In the complex number $a+bi$, a is the _____ and bi is the imaginary part.

Example	Student Practice
1. Simplify.	**2.** Simplify.
(a) $\sqrt{-36}$	**(a)** $\sqrt{-41}$
For all positive real numbers a, $\sqrt{-a} = \sqrt{-1}\sqrt{a} = i\sqrt{a}$.	
$\sqrt{-36} = \sqrt{-1}\sqrt{36} = i(6) = 6i$	
(b) $\sqrt{-17}$	**(b)** $\sqrt{-25}$
$\sqrt{-17} = \sqrt{-1}\sqrt{17} = i\sqrt{17}$	
3. Multiply. $\sqrt{-16}\cdot\sqrt{-25}$	**4.** Multiply. $\sqrt{-36}\cdot\sqrt{-9}$
Rewrite using the definition $\sqrt{-1} = i$.	
$\left(\sqrt{-16}\right)\left(\sqrt{-25}\right) = \left(i\sqrt{16}\right)\left(i\sqrt{25}\right)$ $= i^2(4)(5)$ $= -1(20) = -20$	

Vocabulary Answers: 1. conjugates 2. imaginary number 3. complex number 4. real part

Copyright © 2013 Pearson Education, Inc.

Example	Student Practice
5. Find the real numbers x and y if $x + 3i\sqrt{7} = -2 + yi$.	**6.** Find the real numbers x and y if $-3 + 3yi = x + 12i\sqrt{2}$.

5. Find the real numbers x and y if
$$x + 3i\sqrt{7} = -2 + yi.$$

Two complex numbers $a + bi$ and $c + di$ are equal if and only if $a = c$ and $b = d$.

The real parts must be equal, so x must be -2. The imaginary parts must also be equal, so y must be $3\sqrt{7}$.

6. Find the real numbers x and y if
$$-3 + 3yi = x + 12i\sqrt{2}.$$

---

7. Subtract. $(6 - 2i) - (3 - 5i)$

Subtract the real parts and subtract the imaginary parts.

$$(6 - 2i) - (3 - 5i)$$
$$= (6 - 3) + \left[-2 - (-5)\right]i$$
$$= 3 + (-2 + 5)i$$
$$= 3 + 3i$$

8. Subtract. $(-5 + 6i) - (-2 + 4i)$

---

9. Multiply. $(7 - 6i)(2 + 3i)$

Use FOIL.

$$(7 - 6i)(2 + 3i)$$
$$= (7)(2) + (7)(3i) - (6i)(2) - (6i)(3i)$$
$$= 14 + 21i - 12i - 18i^2$$

Since $i^2 = -1$, $14 + 21i - 12i - 18i^2$ can be written as $14 + 21i - 12i - 18(-1)$. Simplify this expression.

$$14 + 21i - 12i - 18(-1)$$
$$= 14 + 21i - 12i + 18$$
$$= 32 + 9i$$

Thus, $(7 - 6i)(2 + 3i) = 32 + 9i$.

10. Multiply. $(6 - 5i)(3 - 2i)$

Copyright © 2013 Pearson Education, Inc.

Example	Student Practice
11. Multiply. $3i(4-5i)$	**12.** Multiply. $-4i(3-6i)$

Use the distributive property. Recall that $i^2 = 1$.

$$3i(4-5i) = (3)(4)i + (3)(-5)i^2$$
$$= 12i - 15i^2$$
$$= 12i - 15(-1)$$
$$= 15 + 12i$$

---

13. Evaluate.

(a) i^{36}

Some values for i^n are shown below.

$$i = i \quad i^5 = i \quad i^9 = i$$
$$i^2 = -1 \quad i^6 = -1 \quad i^{10} = -1$$
$$i^3 = -i \quad i^7 = -i \quad i^{11} = -i$$
$$i^4 = +1 \quad i^8 = +1 \quad i^{12} = +1$$

Divide the exponent by 4 since i^4 raised to any power will be 1. Then use the values from the table above to evaluate the remainder.

$$i^{36} = \left(i^4\right)^9 = (1)^9 = 1$$

(b) i^{27}

$$i^{27} = \left(i^{24+3}\right) = \left(i^{24}\right)\left(i^3\right)$$
$$= \left(i^4\right)^6\left(i^3\right)$$
$$= (1)^6(-i)$$
$$= -i$$

14. Evaluate.

(a) i^{22}

(b) i^{64}

Copyright © 2013 Pearson Education, Inc.

Example	Student Practice
15. Divide. $\dfrac{3-2i}{4i}$	**16.** Divide. $\dfrac{6-5i}{4+3i}$

Multiply the numerator and denominator by the conjugate of the denominator. The conjugate of $0+4i$ is $0-4i$, or simply $-4i$.

$$\frac{3-2i}{4i} = \frac{3-2i}{4i} \cdot \frac{-4i}{-4i}$$

$$= \frac{-12i+8i^2}{-16i^2}$$

$$= \frac{-12i+8(-1)}{-16(-1)}$$

$$= \frac{-8-12i}{16}$$

$$= \frac{\cancel{4}(-2-3i)}{\cancel{4}\cdot 4}$$

$$= \frac{-2-3i}{4} \text{ or } -\frac{1}{2}-\frac{3}{4}i$$

Extra Practice

1. Multiply and simplify. Place in i notation before doing any other operations.

$$\left(6+\sqrt{-2}\right)\left(3-\sqrt{-2}\right)$$

2. Evaluate. $i^{82}-i^{31}$

3. Multiply and simplify.
$$(3+2i)(2+i)$$

4. Divide. $\dfrac{6}{5-2i}$

Concept Check

Explain how you would simplify the following. $(3+5i)^2$

Copyright © 2013 Pearson Education, Inc.

Chapter 7 Rational Exponents
7.7 Variation

Vocabulary
direct variation   •   inverse variation   •   combined variation   •   constant of variation

1. _____ is where one variable is a constant multiple of the reciprocal of the other.

2. In the direct variation equation $y = kx$, k is the _____.

3. _____ is where one variable is a constant multiple of the other.

4. If a quantity depends on the variation of two or more variables, it is called joint or _____.

Example	Student Practice
1. The time of a pendulum's period varies directly with the square root of its length. If the pendulum is 1 foot long when the time is 0.2 second, find the time when its length is 4 feet.	**2.** A car's stopping distance varies directly with the square of its speed. A car that is traveling 35 miles per hour can stop in 49 feet. What distance will it take to stop if it is traveling 60 miles per hour?

Let $t =$ the time and $L =$ the length.

This gives us the equation $t = k\sqrt{L}$. Substitute $L = 1$ and $t = 0.2$ into the equation to find k.

$$t = k\sqrt{L}$$
$$(0.2) = k\left(\sqrt{1}\right)$$
$$0.2 = k$$

We can rewrite the equation as $t = 0.2\sqrt{L}$. Find the value of t when $L = 4$.

$$t = 0.2\sqrt{L}$$
$$= 0.2\sqrt{4} = (0.2)(2) = 0.4 \text{ seconds}$$

Vocabulary Answers: 1. inverse variation 2. constant of variation 3. direct variation 4. combined variation

Copyright © 2013 Pearson Education, Inc.

Example	Student Practice
3. If y varies inversely with x and $y = 12$ when $x = 5$, find the value of y when $x = 14$.	**4.** If y varies inversely with x and $y = 12$ when $x = 5$, find the value of y when $x = 14$.

3. Write the equation as $y = \dfrac{k}{x}$. Substitute $y = 12$ and $x = 5$ to find k.

$$12 = \frac{k}{5}$$

$$60 = k$$

Rewrite the equation as $y = \dfrac{60}{x}$. Find the value of y when $x = 14$.

$$y = \frac{60}{x} = \frac{60}{14} = \frac{30}{7}$$

5. The amount of light from a light source varies inversely with the square of the distance from the light source. If an object receives 6.25 lumens when the light source is 8 meters away, how much light will the object receive if the light source is 4 meters away?	**6.** The amount of light from a light source varies inversely with the square of the distance from the light source. If an object receives 9.31 lumens when the light source is 10 meters away, how much light will the object receive if the light source is 7 meters away?

5. Let $L =$ light and $d =$ distance, which gives the equation $L = \dfrac{k}{d^2}$. Substitute $L = 6.25$ and $d = 8$ to find k.

$$6.25 = \frac{k}{8^2}$$

$$400 = k$$

Rewrite the original equation with $k = 400$. Then find L when $d = 4$.

$$L = \frac{400}{d^2} = \frac{400}{4^2} = \frac{400}{16} = 25 \text{ lumens}$$

The check is left to the student.

Copyright © 2013 Pearson Education, Inc.

Example	Student Practice
7. y varies directly with x and z and inversely with d^2. When $x = 7$, $z = 3$, and $d = 4$, the value of y is 20. Find the value of y when $x = 5$, $z = 6$, and $d = 2$.	**8.** y varies directly with m^4 and n and inversely with p^2. When $m = 2$, $n = 3$, and $p = 5$, the value of y is 63. Find the value of y when $m = 1$, $n = 8$, and $p = 3$.

Write the equation as $y = \dfrac{kxz}{d^2}$.

Substitute the known values and solve for k.

$$y = \frac{kxz}{d^2}$$

$$20 = \frac{k(7)(3)}{4^2}$$

$$20 = \frac{21k}{16}$$

$$320 = 21k$$

$$\frac{320}{21} = k$$

Substitute this value of k into the original equation.

$$y = \frac{\frac{320}{21}xz}{d^2} \quad \text{or} \quad y = \frac{320xz}{21d^2}$$

Now find y when $x = 5$, $z = 3$, and $d = 2$.

$$y = \frac{320xz}{21d^2}$$

$$y = \frac{320(5)(6)}{21(2)^2} = \frac{9600}{84} = \frac{800}{7}$$

Thus, the value of y is $\dfrac{800}{7}$.

Copyright © 2013 Pearson Education, Inc.

Extra Practice

1. The distance a spring stretches varies directly with the weight of the object hung on the spring. If a 12-pound weight stretches a spring 15 inches, how far will a 32-pound weight stretch this spring?

2. If y varies inversely with the cube of x, and $y = 50$ when $x = 3$, find y when $x = 7$. Round to the nearest tenth.

3. The speed of a car varies inversely with the amount of time it takes to cover a certain distance. At 50 mph, a car travels a certain distance in 16 seconds. What is the speed of the car that travels the same distance in 10 seconds?

4. y varies directly with x, and inversely with z. $y = 40$ when $x = 6$ and $z = 15$. Find the value of y when $x = 0.5$ and $z = 12$. Round to the nearest tenth.

Concept Check

If y varies directly with the square root of x and $y = 50$ when $x = 5$, explain how you would find the constant of variation k.

Copyright © 2013 Pearson Education, Inc.

MATH COACH

Mastering the skills you need to do well on the test.

Watch the **MATH COACH** videos in MyMathLab® or on You Tube™ while you work the problems below. These helpful hints will help you avoid making common errors on test problems.

Simplifying Expressions with Rational Exponents—
Problem 7   Evaluate. $16^{5/4}$

> **Helpful Hint:** First see if the numerical base can be written in exponent form. If there are two alternate ways to rewrite this number, use the form that involves the smallest possible number as the base and has the largest exponent.

Did you write 16 as 2^4?

Yes _____   No _____

If you answered No, think of the different ways that 16 can be written in exponent form: 16^1, 4^2, or 2^4.

Your choice is most likely between 4^2 and 2^4. The problem works out more easily with the smallest possible base, which is 2. See if you can redo this step correctly.

Do you see that if we use the rule $\left(x^m\right)^n = x^{mn}$, we obtain

the expression $\left(2^4\right)^{5/4} = 2^{4/1 \cdot 5/4}$?
Yes _____   No _____

If you answered No, please review the rule about raising a power to a power and see if you can obtain the right result. Now simplify this expression further and evaluate the resulting exponential expression.

If you answered Problem 7 incorrectly, go back and rework the problem using these suggestions.

Adding and Subtracting Radical Expressions—Problem 12
Combine where possible. $\sqrt{40x} - \sqrt{27x} + 2\sqrt{12x}$

> **Helpful Hint:** Remember to simplify each radical expression first. For this problem, remember to keep the variable x inside the radical.

Did you simplify $\sqrt{40x}$ to $2\sqrt{10x}$? Yes _____ No _____
Did you simplify $\sqrt{27x}$ to $3\sqrt{3x}$? Yes _____ No _____

If you answered No to either question, remember that $\sqrt{40x} = \sqrt{4} \cdot \sqrt{10x}$ and $\sqrt{27x} = \sqrt{9} \cdot \sqrt{3x}$. You can take the square roots of both 4 and 9. Try to do that part of the problem again and then simplify to see if you obtain the same results.

Did you simplify $2\sqrt{12x}$ to $4\sqrt{3x}$? Yes _____ No _____

If you answered No, remember that $2\sqrt{12x} = 2 \cdot \sqrt{4} \cdot \sqrt{3x}$. Go back and see if you can obtain the correct answer to that step. Remember that in your final step, you can only add radical expressions if they have the same radicand.

Now go back and rework the problem using these suggestions.

Copyright © 2013 Pearson Education, Inc.

Rationalizing the Denominator—Problem 18 Rationalize the denominator. $\dfrac{1+2\sqrt{3}}{3-\sqrt{3}}$

Helpful Hint: You will need to multiply both the numerator and the denominator by the conjugate of the denominator. Do this very carefully using the FOIL method.

Did you multiply the numerator and the denominator by $\left(3+\sqrt{3}\right)$?

Yes ____ No ____

If you answered No, review the definition of conjugate. Note that the conjugate of $3-\sqrt{3}$ is $3+\sqrt{3}$.

Did you multiply the binomials in the denominator to get $9+3\sqrt{3}-3\sqrt{3}-\sqrt{9}$?

Yes ____ No ____

Did you multiply the binomials in the numerator to get $3+\sqrt{3}+6\sqrt{3}+2\sqrt{9}$?

Yes ____ No ____

If you answered No to either question, go back over each step of multiplying the binomial expressions using the FOIL method. Be careful in writing which numbers go outside the radical and which numbers go inside the radical. Now simplify each expression.

If you answered Problem 18 incorrectly, go back and rework the problem using these suggestions.

Solving a Radical Equation—Problem 20 Solve and check your solution(s). $5+\sqrt{x+15}=x$

Helpful Hint: Always isolate a radical term on one side of the equation before squaring each side.

Did you isolate the radical term first to obtain the equation $\sqrt{x+15}=x-5$?
Yes ____ No ____

After squaring each side, did you obtain $x+15=x^2-10x+25$?
Yes ____ No ____

If you answered No to either question, review the process for solving a radical equation. Remember that $\left(\sqrt{x+15}\right)^2=x+15$ and $(x-5)^2=x^2-10x+25$. Go back and try to complete these steps again.

Did you collect all like terms on one side to obtain the equation $0=x^2-11x+10$?
Yes ____ No ____

If you answered No, remember to add $-x-15$ to both sides of the equation.

Now factor the quadratic equation and set each factor equal to 0 to find the solution(s). Make sure to check your possible solutions and discard any solution that does not check when substituted back into the original equation.

If you answered Problem 20 incorrectly, go back and rework the problem using these suggestions.

Copyright © 2013 Pearson Education, Inc.

Chapter 8 Quadratic Equations and Inequalities
8.1 Quadratic Equations

Vocabulary

quadratic equation • standard form • square root property • complete the square

1. The _____ states that if $x^2 = a$, then $x = \pm\sqrt{a}$ for all real numbers a.

2. When we _____ we are changing the polynomial to a perfect square trinomial.

3. $ax^2 + bx + c = 0$ is the _____ of a quadratic equation.

4. An equation written in the form $ax^2 + bx + c = 0$, where a, b, and c are real numbers and $a \neq 0$, is called a _____.

Example	Student Practice
1. Solve and check. $x^2 - 36 = 0$	**2.** Solve and check. $x^2 - 25 = 0$

1. Solve and check. $x^2 - 36 = 0$

If we add 36 to each side, we have $x^2 = 36$.

$x = \pm\sqrt{36}$
$x = \pm 6$

The two roots are 6 and −6.

Check.

$(6)^2 - 36 \overset{?}{=} 0 \quad (-6)^2 - 36 \overset{?}{=} 0$

$36 - 36 \overset{?}{=} 0 \qquad 36 - 36 \overset{?}{=} 0$

$0 = 0 \qquad\qquad 0 = 0$

2. Solve and check. $x^2 - 25 = 0$

---

3. Solve. $x^2 = 48$

$x = \pm\sqrt{48}$
$x = \pm\sqrt{16 \cdot 3} = \pm 4\sqrt{3}$

The two roots are $4\sqrt{3}$ and $-4\sqrt{3}$.

4. Solve. $x^2 = 24$

Vocabulary Answers: 1. square root property 2. complete the square 3. standard form 4. quadratic equation

Copyright © 2013 Pearson Education, Inc.

Example	Student Practice
5. Solve and check. $3x^2 + 2 = 77$	**6.** Solve and check. $6x^2 - 3 = 93$

$$3x^2 + 2 = 77$$
$$3x^2 = 75$$
$$x^2 = 25$$
$$x = \pm\sqrt{25}$$
$$x = \pm 5$$

The roots are 5 and -5. The check is left to the student.

---

7. Solve and check. $4x^2 = -16$	**8.** Solve and check. $2x^2 = -50$

$$4x^2 = -16$$
$$x^2 = -4$$
$$x = \pm\sqrt{-4}$$
$$x = \pm 2i$$

The roots are $2i$ and $-2i$.

Check.

$$4(2i)^2 \overset{?}{=} -16 \qquad 4(-2i)^2 \overset{?}{=} -16$$
$$4(-4) \overset{?}{=} -16 \qquad 4(-4) \overset{?}{=} -16$$
$$-16 = -16 \qquad -16 = -16$$

---

9. Solve. $(4x-1)^2 = 5$	**10.** Solve. $(3x+8)^2 = 6$

$$(4x-1)^2 = 5$$
$$4x - 1 = \pm\sqrt{5}$$
$$4x = 1 \pm \sqrt{5}$$
$$x = \frac{1 \pm \sqrt{5}}{4}$$

The roots are $\dfrac{1+\sqrt{5}}{4}$ and $\dfrac{1-\sqrt{5}}{4}$.

Copyright © 2013 Pearson Education, Inc.

Example	Student Practice
11. Solve by completing the square and check. $x^2 + 6x + 1 = 0$	**12.** Solve by completing the square and check. $x^2 + 4x + 1 = 0$

First rewrite the equation in the form $ax^2 + bx = c$. Thus, we obtain $x^2 + 6x = -1$.

Next verify that the coefficient of the quadratic term (x^2) is 1. We want to add a constant term to $x^2 + 6x$ so that we get a perfect square trinomial. We do this by taking half the coefficient of x and squaring it.

$$\left(\frac{6}{2}\right)^2 = 3^2 = 9$$

Adding 9 to $x^2 + 6x$ gives the perfect square trinomial $x^2 + 6x + 9$, which we factor as $(x + 3)^2$. Add 9 to both the left and right side of our equation. We now have $x^2 + 6x + 9 = -1 + 9$.

Factor the left side then use the square root property.
$$(x + 3)^2 = 8$$
$$x + 3 = \pm\sqrt{8}$$
$$x + 3 = \pm 2\sqrt{2}$$

Next we solve for x by subtracting 3 from each side of the equation.
$$x = -3 \pm 2\sqrt{2}$$

The roots are $-3 + 2\sqrt{2}$ and $-3 - 2\sqrt{2}$.

The check is left to the student. Be sure to check the solution in the original equation, and not the perfect square trinomial.

Copyright © 2013 Pearson Education, Inc.

Example	Student Practice
13. Solve by completing the square. $3x^2 - 8x + 1 = 0$	**14.** Solve by completing the square. $5x^2 - 4x - 3 = 0$

$3x^2 - 8x + 1 = 0$

$3x^2 - 8x = -1$

Divide each term by 3 so that the coefficient of the quadratic term is 1.

$$\frac{3x^2}{3} - \frac{8x}{3} = -\frac{1}{3}$$

$$x^2 - \frac{8}{3}x + \frac{16}{9} = -\frac{1}{3} + \frac{16}{9}$$

$$\left(x - \frac{4}{3}\right)^2 = \frac{13}{9}$$

$$x - \frac{4}{3} = \pm\sqrt{\frac{13}{9}}$$

$$x - \frac{4}{3} = \pm\frac{\sqrt{13}}{3}$$

$$x = \frac{4 \pm \sqrt{13}}{3}$$

Extra Practice

1. Solve the equation by using the square root property. Express any complex numbers using i notation. $x^2 + 64 = 0$

2. Solve the equation by using the square root property. Express any complex numbers using i notation. $(3x+1)^2 = 15$

3. Solve the equation by completing the square. Express any complex numbers using i notation. $x^2 + 12x + 35 = 0$

4. Solve the equation by completing the square. Express any complex numbers using i notation. $4x^2 + 3 = x$

Concept Check

Explain how you would decide what to add to each side of the equation to complete the square for the following equation. $x^2 + x = 1$

Copyright © 2013 Pearson Education, Inc.

Name: _____   Date: _____

Instructor: _____   Section: _____

Chapter 8 Quadratic Equations and Inequalities
8.2 The Quadratic Formula and Solutions to Quadratic Equations

Vocabulary
quadratic formula　•　standard form　•　discriminant　•　complex number

1.　The expression $b^2 - 4ac$ is called the _____.

2.　The _____ says that for all equations $ax^2 + bx + c = 0$, $x = \dfrac{-b \pm \sqrt{b^2 - 4ac}}{2a}$.

Example	**Student Practice**
1. Solve by using the quadratic formula. $x^2 + 8x = -3$ The standard form is $x^2 + 8x + 3 = 0$. We substitute $a = 1$, $b = 8$, and $c = 3$. $x = \dfrac{-b \pm \sqrt{b^2 - 4ac}}{2a}$ $x = \dfrac{-8 \pm \sqrt{8^2 - 4(1)(3)}}{2(1)}$ $x = \dfrac{-8 \pm 2\sqrt{13}}{2}$ $x = -4 \pm \sqrt{13}$	**2.** Solve by using the quadratic formula. $x^2 + 4x = -7 - 8x$
3. Solve by using the quadratic formula. $2x^2 - 48 = 0$ The standard form is $2x^2 - 0x - 48 = 0$. Therefore, $a = 2$, $b = 0$, and $c = -48$. $x = \dfrac{-0 \pm \sqrt{(0)^2 - 4(2)(-48)}}{2(2)}$ $x = \dfrac{\pm 8\sqrt{6}}{4}$ $x = \pm 2\sqrt{6}$	**4.** Solve by using the quadratic formula. $4x^2 - 60 = 0$

Vocabulary Answers: 1. discriminant 2. quadratic formula

Copyright © 2013 Pearson Education, Inc.

Example	Student Practice

5. A small company that manufactures canoes makes a daily profit p according to the equation $p = -100x^2 + 3400x - 26{,}196$, where p is measured in dollars and x is the number of canoes made per day. Find the number of canoes that must be made each day to produce a zero profit for the company. Round your answer to the nearest whole number.

Since $p = 0$, the equation is $0 = -100x^2 + 3400x - 26{,}196$. Thus, $a = -100$, $b = 3400$, and $c = -26{,}196$.

$$x = \frac{-3400 \pm \sqrt{3400^2 - 4(-100)(-26{,}196)}}{2(-100)}$$

$$x = \frac{-3400 \pm \sqrt{1{,}081{,}600}}{-200}$$

$$x = \frac{-3400 \pm 1040}{-200}$$

We now obtain two answers.

$$x = \frac{-3400 + 1040}{-200}$$

$$x = \frac{-2360}{-200} = 11.8 \approx 12$$

$$x = \frac{-3400 - 1040}{-200}$$

$$x = \frac{-4440}{-200} = 22.2 \approx 22$$

A zero profit is obtained when approximately 12 canoes are produced or when approximately 22 canoes are produced.

6. A company that manufactures guitars makes a daily profit p according to the equation $p = -100x^2 + 3024x - 20{,}240$, where p is measured in dollars and x is the number of guitars made per day. Find the number of guitars that must be made each day to produce a zero profit for the company. Round your answer to the nearest whole number.

Copyright © 2013 Pearson Education, Inc.

Example	Student Practice
7. Solve by using the quadratic formula. $$\frac{2x}{x+2} = 1 - \frac{3}{x+4}$$ The LCD is $(x+2)(x+4)$. $$\frac{2x}{x+2} = 1 - \frac{3}{x+4}$$ $$2x(x+4) = (x+2)(x+4) - 3(x+2)$$ $$x^2 + 5x - 2 = 0$$ $$x = \frac{-5 \pm \sqrt{5^2 - 4(1)(-2)}}{2(1)}$$ $$x = \frac{-5 \pm \sqrt{33}}{2}$$	**8.** Solve by using the quadratic formula. $$\frac{6}{x} + \frac{x}{x-3} = -\frac{4}{5}$$
9. Solve and simplify your answer. $$8x^2 - 4x + 1 = 0$$ $$x = \frac{-(-4) \pm \sqrt{(-4)^2 - 4(8)(1)}}{2(8)}$$ $$x = \frac{4 \pm 4i}{16}$$ $$x = \frac{1 \pm i}{4}$$	**10.** Solve by using the quadratic formula. $$6x^2 + 4x + 3 = 0$$
11. What type of solutions does the equation $2x^2 - 9x - 35 = 0$ have? Do not solve the equation. $a = 2,\ b = -9,$ and $c = -35.$ $$b^2 - 4ac = \left(-9^2\right) - 4(2)(-35) = 361$$ Since the discriminant is positive, the equation has two real roots. 361 is a perfect square. The equation has two different rational solutions.	**12.** Use the discriminant to find what type of solutions the equation $2x^2 - 5x + 43 = 0$ has. Do not solve the equation.

Copyright © 2013 Pearson Education, Inc.

Example	Student Practice
13. Find a quadratic equation whose roots are 5 and −2. $x = 5 \qquad x = -2$ $x - 5 = 0 \quad x + 2 = 0$ $(x - 5)(x + 2) = 0$ $x^2 - 3x - 10 = 0$	**14.** Find a quadratic equation whose roots are 3 and 4.
15. Find a quadratic equation those solutions are $3i$ and $-3i$. $x - 3i = 0 \quad \text{and} \quad x + 3i = 0$ $(x - 3i)(x + 3i) = 0$ $x^2 + 3ix - 3ix - 9i^2 = 0$ $x^2 + 9 = 0$	**16.** Find a quadratic equation whose solutions are $3i\sqrt{5}$ and $-3i\sqrt{5}$

Extra Practice

1. Solve by the quadratic formula. Simplify your answers. $3x^2 + 12x + 7 = 0$

2. Simplify, then solve by the quadratic formula. Simplify your answers. Use i notation for nonreal complex numbers.
$$\frac{1}{x} - \frac{2}{x+2} = \frac{1}{5}$$

3. Use the discriminant to find what type of solutions the following equation has. Do not solve the equation.
$x^2 - 3(2x - 3) = 0$

4. Write a quadratic equation having the given solutions -3 and $-\dfrac{1}{2}$.

Concept Check

Explain how you would determine if $2x^2 - 6x + 3 = 3$ has two rational, two irrational, one rational, or two nonreal complex solutions.

Copyright © 2013 Pearson Education, Inc.

Chapter 8 Quadratic Equations and Inequalities
8.3 Equations That Can Be Transformed into Quadratic Form

Vocabulary
quadratic in form   •   linear term   •   standard form   •   substitution

1. Before making a substitution and then factoring, we first must put the equation into _____.

2. An equation is _____ if we can substitute a linear term for the variable raised to the lowest power and get an equation of the form $ay^2 + by + c = 0$.

Example	Student Practice
1. Solve. $x^4 - 13x^2 + 36 = 0$	**2.** Solve. $x^4 + 9x^2 - 400 = 0$

1. Solve. $x^4 - 13x^2 + 36 = 0$

Let $y = x^2$. Then $y^2 = x^4$. Substitute these into the equation.

$y^2 - 13y + 36 = 0$

Factor the equation and then solve for x.

$(y - 4)(y - 9) = 0$
$y - 4 = 0$ or $y - 9 = 0$
$\quad y = 4 \qquad\qquad y = 9$

Replace y with x^2.

$x^2 = 4 \qquad$ or $\qquad x^2 = 9$
$\quad x = \pm\sqrt{4} \qquad\qquad x = \pm\sqrt{9}$
$\quad x = \pm 2 \qquad\qquad\quad x = \pm 3$

Thus, there are four solutions to the original equation: $x = +2$, $x = -2$, $x = +3$, and $x = -3$. Check these values to verify that they are solutions.

The check is left to the student.

Vocabulary Answers: 1. standard form 2. quadratic in form

Copyright © 2013 Pearson Education, Inc.

Example	Student Practice
3. Solve for all real roots. $2x^6 - x^3 - 6 = 0$	**4.** Solve for all real roots. $3x^6 - 83x^3 + 54 = 0$

Let $y = x^3$. Then $y^2 = x^6$. Thus, we have the following:

$$2y^2 - y - 6 = 0$$
$$(2y + 3)(y - 2) = 0$$
$$2y + 3 = 0 \qquad \text{or} \quad y - 2 = 0$$
$$y = -\frac{3}{2} \qquad\qquad y = 2$$
$$x^3 = -\frac{3}{2} \qquad\qquad x^3 = 2$$
$$x = \sqrt[3]{-\frac{3}{2}} \qquad x = \sqrt[3]{2}$$
$$x = \frac{\sqrt[3]{-12}}{2}$$

The check is left to the student.

Example	Student Practice
5. Solve and check your solutions. $x^{2/3} - 3x^{1/3} + 2 = 0$	**6.** Solve and check your solutions. $x^{2/3} - 65x^{1/3} + 64 = 0$

Let $y = x^{1/3}$. Then $y^2 = x^{2/3}$.

$$y^2 - 3y + 2 = 0$$
$$(y - 2)(y - 1) = 0$$
$$y - 2 = 0 \quad \text{or} \quad y - 1 = 0$$
$$y = 2 \qquad\qquad y = 1$$
$$x^{1/3} = 2 \qquad\qquad x^{1/3} = 1$$
$$\left(x^{1/3}\right)^3 = (2)^3 \quad \left(x^{1/3}\right)^3 = (1)^3$$
$$x = 8 \qquad\qquad x = 1$$

The check is left to the student.

Copyright © 2013 Pearson Education, Inc.

Example	Student Practice
7. Solve and check your solutions. $2x^{1/2} = 5x^{1/4} + 12$	**8.** Solve and check your solutions. $2x^{1/2} = 11x^{1/4} + 15$

7. Solve and check your solutions.

$2x^{1/2} = 5x^{1/4} + 12$

This equation in standard form is
$2x^{1/2} - 5x^{1/4} - 12 = 0$. Let $y = x^{1/4}$,
making $y^2 = x^{1/2}$. Then solve.

$$2y^2 - 5y - 12 = 0$$
$$(2y+3)(y-4) = 0$$
$$2y + 3 = 0 \qquad \text{or} \quad y - 4 = 0$$
$$y = -\frac{3}{2} \qquad\qquad y = 4$$
$$x^{1/4} = -\frac{3}{2} \qquad\qquad x^{1/4} = 4$$
$$\left(x^{1/4}\right)^4 = \left(-\frac{3}{2}\right)^4 \qquad \left(x^{1/4}\right)^4 = (4)^4$$
$$x = \frac{81}{16} \qquad\qquad x = 256$$

Check by substituting the solutions into
the original equation.

$x = \frac{81}{16}$

$$2\left(\frac{81}{16}\right)^{1/2} - 5\left(\frac{81}{16}\right)^{1/4} - 12 \overset{?}{=} 0$$
$$\frac{9}{2} - \frac{15}{2} - 12 \overset{?}{=} 0$$
$$-15 \neq 0$$

$x = 256$

$$2(256)^{1/2} - 5(256)^{1/4} - 12 \overset{?}{=} 0$$
$$32 - 20 - 12 \overset{?}{=} 0$$
$$0 = 0$$

$\frac{81}{16}$ is extraneous and not a valid
solution. The only valid solution is 256.

Copyright © 2013 Pearson Education, Inc.

Example	Student Practice
9. Solve and check your solutions. $x^{-2} = 5x^{-1} + 14$	**10.** Solve and check your solutions. $x^{-2} + x^{-1} = 12$

Write the equation in standard form. Let $y = x^{-1}$, making $y^2 = x^{-2}$. Then solve.

$$x^{-2} = 5x^{-1} + 14$$
$$x^{-2} - 5x^{-1} - 14 = 0$$
$$y^2 - 5y - 14 = 0$$
$$(y-7)(y+2) = 0$$
$$y - 7 = 0 \quad \text{or} \quad y + 2 = 0$$
$$y = 7 \qquad y = -2$$
$$x^{-1} = 7 \qquad x^{-1} = -2$$
$$x = \frac{1}{7} \qquad x = -\frac{1}{2}$$

The check is left to the student.

Extra Practice

1. Solve. Express any nonreal complex numbers with i notation.
$$x^4 + 10x^2 + 21 = 0$$

2. Solve for all real roots. $x^6 + 27x^3 = 0$

3. Solve for all real roots.
$$6x^{2/5} + 18x^{1/5} + 12 = 0$$

4. Solve for all real roots. $3x^{-2} + 3x^{-1} = 168$

Concept Check

Explain how you would solve the following. $x^8 - 6x^4 = 0$

Copyright © 2013 Pearson Education, Inc.

Name: _____   Date: _____

Instructor: _____   Section: _____

Chapter 8 Quadratic Equations and Inequalities
8.4 Formulas and Applications

Vocabulary
hypotenuse • leg • Pythagorean Theorem • area • surface area

1. The _____ states that if c is the length of the longest side of a right triangle and a and b are the lengths of the other two sides, then $a^2 + b^2 = c^2$.

2. The longest side of a right triangle is called the _____.

Example	Student Practice
1. The surface area of a sphere is given by $A = 4\pi r^2$. Solve this equation for r. (You do not need to rationalize the denominator.) $$A = 4\pi r^2$$ $$\frac{A}{4\pi} = r^2$$ $$\pm\sqrt{\frac{A}{4\pi}} = r$$ $$\pm\frac{1}{2}\sqrt{\frac{A}{\pi}} = r$$ Since the radius of a sphere must be a positive value, we use only the principal root. $r = \frac{1}{2}\sqrt{\frac{A}{\pi}}$	**2.** The volume of a sphere is given by $V = \frac{4}{3}\pi r^3$. Solve this equation for r. (You do not need to rationalize the denominator.)
3. Solve for y. $y^2 - 2yz - 15z^2 = 0$ Factor, then set each factor equal to 0. $$y^2 - 2yz - 15z^2 = 0$$ $$(y+3z)(y-5z) = 0$$ $$y+3z = 0 \quad \text{or} \quad y-5z = 0$$ $$y = -3z \qquad\qquad y = 5z$$	**4.** Solve for x. $x^2 - 3xy - 10y^2 = 0$

Vocabulary Answers: 1. Pythagorean Theorem 2. hypotenuse

Copyright © 2013 Pearson Education, Inc.

Example	Student Practice
5. Solve for x. $2x^2 + 3wx - 4z = 0$	**6.** Solve for x. $4x^2 - 6yx + 9z = 0$

5. Solve for x. $2x^2 + 3wx - 4z = 0$

We use the quadratic formula where the variable is considered to be x, and w and z are considered constants. Thus, $a = 2$, $b = 3w$, and $c = -4z$.

$$x = \frac{-b \pm \sqrt{b^2 - 4ac}}{2a}$$

$$x = \frac{-3w \pm \sqrt{(3w)^2 - 4(2)(-4z)}}{2(2)}$$

$$x = \frac{-3w \pm \sqrt{9w^2 + 32z}}{4}$$

---

7. Complete parts **(a)** and **(b)**.

(a) Solve the Pythagorean Theorem $a^2 + b^2 = c^2$ for a.

$$a^2 + b^2 = c^2$$
$$a^2 = c^2 - b^2$$
$$a = \pm\sqrt{c^2 - b^2}$$

a, b, and c must be positive numbers because they represent lengths, only use the positive root, $a = \sqrt{c^2 - b^2}$.

(b) Find the value of a if $c = 13$ and $b = 5$.

$$a = \sqrt{c^2 - b^2}$$
$$a = \sqrt{(13)^2 - (5)^2}$$
$$a = \sqrt{144}$$
$$a = 12$$

8. Complete parts **(a)** and **(b)**.

(a) Solve the Pythagorean Theorem $a^2 + b^2 = c^2$ for c.

(b) Find the value of c if $a = 15$ and $b = 20$.

Copyright © 2013 Pearson Education, Inc.

Example	Student Practice
9. The perimeter of a triangular piece of land is 12 miles. One leg of the triangle is 1 mile longer than the other leg. Find the length of each boundary of the land if the triangle is a right triangle.	**10.** The perimeter of a triangular piece of land is 60 miles. One leg of the triangle is 4 miles longer than twice the other leg. Find the length of each boundary of the land if the triangle is a right triangle.

We are given that the perimeter is 12 miles, so $x + (x+1) + c = 12$.
Thus, $c = -2x + 11$.

By the Pythagorean Theorem,
$x^2 + (x+1)^2 = (-2x+11)^2$.

$$x^2 + (x+1)^2 = (-2x+11)^2$$
$$x^2 + x^2 + 2x + 1 = 4x^2 - 44x + 121$$
$$0 = 2x^2 - 46x + 120$$
$$0 = x^2 - 23x + 60$$

Use the quadratic formula to solve.
$$x = \frac{-(23) \pm \sqrt{(-23)^2 - 4(1)(60)}}{2(1)}$$

$$x = \frac{23 \pm \sqrt{289}}{2}$$

$$x = \frac{23 \pm 17}{2}$$

$$x = \frac{40}{2} = 20 \text{ or } x = \frac{6}{2} = 3.$$

20 is too large, so the only answer that makes sense is $x = 3$.

Thus, the sides of the triangle are $x = 3$, $x+1 = 4$, and $-2x+11 = -2(3) + 11 = 5$.

The longest boundary of this triangular piece of land is 5 miles. The other two boundaries are 4 miles and 3 miles.

Copyright © 2013 Pearson Education, Inc.

Example	Student Practice
11. A triangular sign marks the edge of the rocks in Rockport Harbor. The sign has an area of 35 square meters. Find the base and altitude of this triangular sign if the base is 3 meters shorter than the altitude.	**12.** The length of a rectangle is 2 yards shorter than twice the width. The area of the rectangle is 84 square yards. Find the dimensions of the rectangle.

The area of the triangle is given by $A = \frac{1}{2}ab$. Let $x =$ the length in meters of the altitude. Then $x - 3 =$ the length in meters of the base. Solve for x.

$$35 = \frac{1}{2}x(x-3)$$
$$0 = x^2 - 3x - 70$$
$$0 = (x-10)(x+7)$$
$$x = 10 \text{ or } x = -7$$

We disregard -7. Thus altitude $= x = 10$ meters and base $= x - 3 = 7$ meters. The check is left to the student.

Extra Practice

1. Solve $s = \frac{1}{2}gt^2$ for t. Assume that all other variables are nonzero.

2. Solve $(a+2)x^2 - 7x + 3y = 0$ for x. Assume that all other variables are nonzero.

3. $c = 10$ and $b = 3a$; use the Pythagorean Theorem to find a and b.

4. Andrew drove at a constant speed on a dirt road for 75 miles. He then traveled 20 mph faster on a paved road for 180 miles. If he drove for 7 hours, find the car's speed for each part of the trip.

Concept Check

In a right triangle the hypotenuse measures 12 meters. One of the legs of the triangle is three times as long as the other. Explain how you would find the length of each leg.

Copyright © 2013 Pearson Education, Inc.

Chapter 8 Quadratic Equations and Inequalities
8.5 Quadratic Functions

Vocabulary
vertex   •   quadratic function

1. A _____ is a function of the form $f(x) = ax^2 + bx + c$.

2. The _____ is the lowest point on a parabola opening upward or the highest point on a parabola opening downward.

Example	Student Practice
1. Find the coordinates of the vertex and the intercepts of the quadratic function $f(x) = x^2 - 8x + 15$.	**2.** Find the coordinates of the vertex and the intercepts of the quadratic function $g(x) = x^2 - 7x + 12$.

The vertex occurs at $\dfrac{-b}{2a} = \dfrac{-(-8)}{2(1)} = 4$.

To find the y-coordinate, find $f(4)$.

$f(4) = 4^2 - 8(4) + 15$
$ = 16 - 32 + 15 = -1$

The vertex is $(4, -1)$. The y-intercept is at $f(0)$.

$f(0) = 0^2 - 8(0) + 15 = 15$

The y-intercept is $(0, 15)$. The x-intercepts occur when $x^2 - 8x + 15 = 0$. Solve for x.

$(x-5)(x-3) = 0$

$x - 5 = 0 \quad x - 3 = 0$
$ x = 5 \qquad x = 3$

The x-intercepts are $(5, 0)$ and $(3, 0)$.

Vocabulary Answers: 1. quadratic function 2. vertex

Copyright © 2013 Pearson Education, Inc.

Example	Student Practice

Example

3. Find the vertex and the intercepts, and then graph the function $f(x) = x^2 + 2x - 4$.

Since $a > 0$, the parabola opens upward. Find the vertex.

$$x = \frac{-b}{2a} = \frac{-2}{2(1)} = \frac{-2}{2} = -1$$

$$f(-1) = (-1)^2 + 2(-1) - 4 = -5$$

The vertex is $(-1, -5)$. The y-intercept is at $f(0)$.

$$f(0) = (0)^2 + 2(0) - 4 = -4$$

The y-intercept is $(0, -4)$. The x-intercepts occur when $f(x) = 0$. We use the quadratic formula.

$$x = \frac{-b \pm \sqrt{b^2 - 4ac}}{2a} = \frac{-2 \pm \sqrt{2^2 - 4(1)(-4)}}{2(1)}$$
$$= -1 \pm \sqrt{5}$$

The x-intercepts are approximately $(-3.2, 0)$ and $(1.2, 0)$. The vertex is $(-1, -5)$; the y-intercept is $(0, -4)$; and the x-intercepts are approximately $(-3.2, 0)$ and $(1.2, 0)$. Connect these points by a smooth curve to graph the parabola.

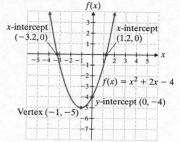

Student Practice

4. Find the vertex and the intercepts, and then graph the function $f(x) = x^2 - 4x + 1$.

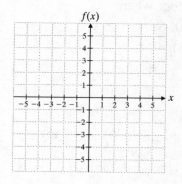

Copyright © 2013 Pearson Education, Inc.

Example	Student Practice
5. Find the vertex and the intercepts, and then graph the function $f(x) = -2x^2 + 4x - 3$.	**6.** Find the vertex and the intercepts, and then graph the function $f(x) = -3x^2 - 6x - 6$.

Since $a < 0$, the parabola opens downward. Find the vertex.

$$x = \frac{-4}{2(-2)} = \frac{-4}{-4} = 1$$

$$f(1) = -2(1)^2 + 4(1) - 3 = -1$$

The vertex is $(1, -1)$. The y-intercept is at $f(0)$. $f(0) = -2(0)^2 + 4(0) - 3$ $= -3$. The y-intercept is $(0, -3)$. The x-intercepts occur when $f(x) = 0$. We use the quadratic formula.

$$x = \frac{-4 \pm \sqrt{4^2 - 4(-2)(-3)}}{2(-2)}$$

$$= \frac{-4 \pm \sqrt{-8}}{-4}$$

Because $\sqrt{-8}$ yields an imaginary number, there are no x-intercepts for the graph of the function. We will look for three additional points. We try $f(2)$, $f(3)$, and $f(-1)$ and get the points $(2, -3)$, $(3, -9)$, and $(-1, -9)$.

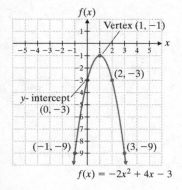

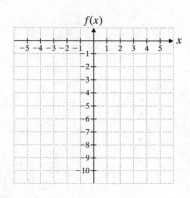

Copyright © 2013 Pearson Education, Inc.

Extra Practice

1. Find the coordinates of the vertex and the intercepts of $g(x) = x^2 + 5x - 6$. When necessary, approximate x-intercepts to the nearest tenth.

2. Find the coordinates of the vertex and the intercepts of $f(x) = 5x^2 + 17x - 12$. When necessary, approximate x-intercepts to the nearest tenth.

3. Find the vertex, the y-intercept, and the x-intercepts (if any exist), and then graph the function $r(x) = 3x^2 - 2x + 1$.

4. Find the vertex, the y-intercept, and the x-intercepts (if any exist), and then graph the function $f(x) = -x^2 + 2x - 1$.

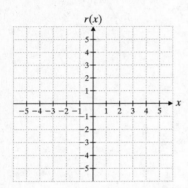

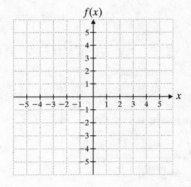

Concept Check

Explain how you would find the vertex of $f(x) = 4x^2 - 9x - 5$.

Copyright © 2013 Pearson Education, Inc.

Name: _____   Date: _____
Instructor: _____   Section: _____

Chapter 8 Quadratic Equations and Inequalities
8.6 Quadratic Inequalities in One Variable

Vocabulary
quadratic inequality   •   boundary points

1. The two points where the expression of a quadratic inequality is equal to zero are called _____.

2. A _____ has the form $ax^2 + bx + c < 0$ (or replace < by >, ≤, or ≥), where a, b, and c are real numbers and $a \neq 0$.

Example	**Student Practice**
1. Solve and graph $x^2 - 10x + 24 > 0$.	**2.** Solve and graph $x^2 - x - 12 > 0$.

Replace the inequality symbol with an equals sign and solve the resulting equation.

$$x^2 - 10x + 24 = 0$$
$$(x - 4)(x - 6) = 0$$
$$x - 4 = 0 \text{ or } x - 6 = 0$$
$$x = 4 \qquad x = 6$$

We use the boundary points to separate the number line into distinct regions.

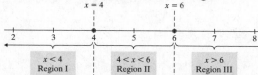

Evaluate the quadratic expression at a test point in each of the regions. Pick the test point $x = 1$ and get $15 > 0$. Pick the test point $x = 5$ and get $-1 < 0$. Pick the test point $x = 7$ and get $3 > 0$. Thus, $x^2 - 10x + 24 > 0$ when $x < 4$ or when $x > 6$.

Vocabulary Answers: 1. boundary points 2. quadratic inequality

Example	Student Practice
3. Solve and graph $2x^2 + x - 6 \le 0$.	**4.** Solve and graph $2x^2 + x - 15 \le 0$.

We replace the inequality symbol with an equals sign and solve the resulting equation.

$$2x^2 + x - 6 = 0$$
$$(2x - 3)(x + 2) = 0$$

Apply the zero factor property.

$$2x - 3 = 0 \quad \text{or} \quad x + 2 = 0$$
$$2x = 3 \qquad\qquad x = -2$$
$$x = \frac{3}{2} = 1.5$$

We use the boundary points to separate the number line into distinct regions. The boundary points are $x = -2$ and $x = 1.5$. Now we arbitrarily pick a test point in each region.

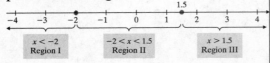

We will pick 3 values of x for the polynomial $2x^2 + x - 6$. First we need an x- value less than -2. We pick $x = -3$ and get $9 > 0$. Next we need an x- value between -2 and 1.5. We pick $x = 0$ and get $-6 < 0$. Finally we need an x- value greater than 1.5. We pick $x = 2$ and get $4 > 0$.

Since our inequality is $\le$ and not just $<$, we need to include the boundary points. Thus, $2x^2 + x - 6 \le 0$ when $-2 \le x \le 1.5$.

Copyright © 2013 Pearson Education, Inc.

Example	Student Practice
5. Solve and graph $x^2 + 4x > 6$. Round your answer to the nearest tenth.	**6.** Solve and graph $x^2 + 4x > 1$. Round your answer to the nearest tenth.

First we write $x^2 + 4x - 6 > 0$. Because we cannot factor $x^2 + 4x - 6$, we use the quadratic formula to find the boundary points.

$$x = \frac{-4 \pm \sqrt{4^2 - 4(1)(-6)}}{2(1)}$$

$$= \frac{-4 \pm \sqrt{40}}{2}$$

$$= -2 \pm \sqrt{10}$$

$-2 + \sqrt{10} \approx 1.2$ or $-2 - \sqrt{10} \approx -5.2$

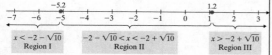

We will see where $x^2 + 4x - 6 > 0$. Test $x = -6$.

$$(-6)^2 + 4(-6) - 6 = 36 - 24 - 6 = 6 > 0$$

Test $x = 0$.

$$(0)^2 + 4(0) - 9 = 0 + 0 - 6 = -6 < 0$$

Test $x = 2$.

$$(2)^2 + 4(2) - 6 = 4 + 8 - 6 = 6 > 0$$

Our answer is $x < -5.2$ or $x > 1.2$.

Copyright © 2013 Pearson Education, Inc.

Extra Practice

1. Solve and graph.

$$2x^2 - 3x - 5 \le 0$$

2. Solve.

$$x^2 - 12x > 64$$

3. Solve.

$$x^2 + 81 \ge 18x$$

4. Solve the following quadratic equation if possible. Round your answers to the nearest tenth.

$$7x^2 \ge 5x^2 - 8$$

Concept Check

Explain what happens when you solve the inequality $x^2 + 2x + 8 > 0$.

Copyright © 2013 Pearson Education, Inc.

MATH COACH

Mastering the skills you need to do well on the test.

Watch the **MATH COACH** videos in MyMathLab® or on You[Tube]™ while you work the problems below. These helpful hints will help you avoid making common errors on test problems.

Solving Quadratic Equations Involving Fractions—Problem 6

Solve the quadratic equation. $\dfrac{2x}{2x+1} - \dfrac{6}{4x^2-1} = \dfrac{x+1}{2x-1}$

Helpful Hint: If any denominators need to be factored, do that first. Then determine the LCD of all the denominators in the equation. Multiply each term of the equation by the LCD before solving for x.

Did you factor $4x^2-1$ as $(2x+1)(2x-1)$?

Yes _____ No _____

Did you identify the LCD to be $(2x+1)(2x-1)$?

Yes _____ No _____

If you answered No to these questions, review how to factor the difference of two squares and how to find the LCD of polynomial denominators.

Did you multiply the LCD by each term of the equation and remove parentheses to obtain $2x^2-5x-7=0$?

Yes _____ No _____

Did you then use the quadratic formula and substitute for a, b, and c to get $x = \dfrac{-(-5)\pm\sqrt{(-5)^2-4(2)(-7)}}{2(2)}$?

Yes _____ No _____

If you answered No to these questions, remember that your final equation should be in the form $ax^2+bx+c=0$. Then use $a=2$, $b=-5$, and $c=-7$ in the quadratic formula and simplify your result.

If you answered Problem 6 incorrectly, go back and rework the problem using these suggestions.

Solving Equations That Are Quadratic in Form—Problem 9

Solve for any valid real roots. $x^4-11x^2+18=0$

Helpful Hint: Let $y=x^2$ and then let $y^2=x^4$. Write the new quadratic equation after these replacements have been made.

After making the necessary replacements, did you obtain the equation $y^2-11y+18=0$? Yes _____ No _____

Did you solve the quadratic equation using any method to result in $y=9$ and $y=2$? Yes _____ No _____

If you answered No to these questions, stop and complete these steps again.

If $y=9$ and $y=2$, can you conclude that $x^2=9$ and $x^2=2$? Yes _____ No _____

If you take the square root of each side of each equation, can you obtain $x=\pm3$ and $x=\pm\sqrt{2}$? Yes _____ No _____

If you answered No to these questions, remember that when you take the square root of each side of the equation there are two sign possibilities. Note that your final solution should consist of four values for x.

Now go back and rework the problem using these suggestions.

Copyright © 2013 Pearson Education, Inc.

Solving a Quadratic Equation with Several Variables—Problem 13

Solve for the variable specified. $5y^2 + 2by + 6w = 0$; for y

> **Helpful Hint:** Think of the equation being written as $ay^2 + by + c = 0$. The quantities for a, b, or c may contain variables. Use the quadratic formula to solve for y.

Did you determine that $a = 5$, $b = 2b$, and $c = 6w$?
Yes _____ No _____
Did you substitute these values into the quadratic formula

and simplify to obtain $y = \dfrac{-2b \pm \sqrt{4b^2 - 120w}}{10}$?

Yes _____ No _____

If you answered No to these questions, review the Helpful Hint again to make sure that you find the correct values for a, b, and c. Carefully substitute these values into the quadratic formula and simplify.

Were you able to simplify the radical expression by factoring

out a 4 to obtain $\sqrt{4(b^2 - 30w)}$? Yes _____ No _____

Did you simplify this expression further to get

$2\sqrt{b^2 - 30w}$? Yes _____ No _____

If you answered No to these questions, remember to always simplify radicals whenever possible. Make sure to write your solution as a simplified rational expression.

Now go back and rework the problem using these suggestions.

Find the Vertex, Intercepts and Graph of a Quadratic Function—Problem 17

Find the vertex and the intercepts of $f(x) = -x^2 - 6x - 5$. Then graph the function.

> **Helpful Hint:** When the function is written in $f(x) = ax^2 + bx + c$ form, we can find the vertex using the vertex formula. We can solve for the intercepts using the substitutions $x = 0$ and $f(x) = 0$ to find the unknown coordinates. And, if $a < 0$, the graph is a parabola opening downward.

Do you see that $a = -1$, $b = -6$, and $c = -5$?
Yes _____ No _____

Did you use the vertex formula to discover that the vertex point has an x-coordinate of -3? Yes _____ No _____

If you answered No to these questions, notice that the function is written in $f(x) = ax^2 + bx + c$ form and review

the vertex formula: $x = \dfrac{-b}{2a}$. Substitute the resulting value

for x into the original function to find the y-coordinate of the vertex point.

To find the y-intercept, did you substitute 0 for x into the original function to find the value for y?
Yes _____ No _____

After letting $f(x) = 0$ and substituting the values for a, b, and c into the quadratic formula, did you get the expression

$x = \dfrac{-(-6) \pm \sqrt{(-6)^2 - 4(-1)(-5)}}{2(-1)}$? Yes _____ No _____

If you answered No to these questions, remember that the y-intercept will be an ordered pair in the form $(0, y)$ or in this case, $(0, f(x))$, and the x-intercept will be an ordered pair in the form $(x, 0)$. Be careful when substituting values for a, b, and c into the quadratic formula and remember to evaluate $\sqrt{16}$ as both 4 and -4. Simplify the expression to find the possible x-values.

Since $a < 0$, the parabola will open downward. Plot the vertex, x-intercept, and y-intercept points and connect these points with a curve to find the graph of the function.

If you answered Problem 17 incorrectly, go back and rework the problem using these suggestions.

Copyright © 2013 Pearson Education, Inc.

Chapter 9 The Conic Sections
9.1 The Distance Formula and the Circle

Vocabulary
conic section   •   distance formula   •   circle   •   radius   •   center

1. A _____ is defined as the set of all points in a plane that are a fixed distance from a point in that plane.

2. The _____ states that the distance between two points (x_1, y_1) and (x_2, y_2) is $d = \sqrt{(x_2 - x_1)^2 + (y_2 - y_1)^2}$.

3. A _____ is a shape formed by slicing a cone with a plane.

4. The point from which the set of points in a circle are a fixed distance from is called the _____.

Example	Student Practice
1. Find the distance between $(3, -4)$ and $(-2, -5)$.	**2.** Find the distance between $(5, 10)$ and $(3, 5)$.

To use the distance formula, we arbitrarily let $(x_1, y_1) = (3, -4)$ and $(x_2, y_2) = (-2, -5)$. Substitute the appropriate values into the formula.

$$d = \sqrt{(x_2 - x_1)^2 + (y_2 - y_1)^2}$$
$$= \sqrt{(-2 - 3)^2 + [-5 - (-4)]^2}$$
$$= \sqrt{(-5)^2 + (-5 + 4)^2}$$
$$= \sqrt{(-5)^2 + (-1)^2}$$
$$= \sqrt{25 + 1}$$
$$= \sqrt{26}$$

Vocabulary Answers: 1. circle 2. distance formula 3. conic section 4. center

Copyright © 2013 Pearson Education, Inc.

Example	Student Practice
3. Find the center and radius of the circle $(x-2)^2+(y-3)^2=25$. Then sketch its graph.	**4.** Find the center and radius of the circle $(x-1)^2+(y-4)^2=16$. Then sketch its graph.

3. Find the center and radius of the circle $(x-2)^2+(y-3)^2=25$. Then sketch its graph.

From the equation of a circle, $(x-h)^2+(y-k)^2=r^2$, we see that $(h,k)=(2,3)$. Thus, the center of the circle is at $(2,3)$. Since $r^2=25$, the radius of the circle is $r=5$.

To graph the circle, start by graphing the center. Then, use the radius to graph the circle. The graph is shown below.

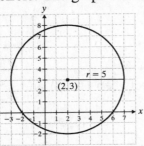

4. Find the center and radius of the circle $(x-1)^2+(y-4)^2=16$. Then sketch its graph.

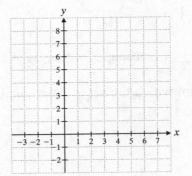

5. Write the equation of the circle with center $(-1,3)$ and radius $\sqrt{5}$. Put your answer in standard form.

We are given that $(h,k)=(-1,3)$ and $r=\sqrt{5}$. Substitute the values into the standard form, $(x-h)^2+(y-k)^2=r^2$.

$$(x-h)^2+(y-k)^2=r^2$$
$$[x-(-1)]^2+(y-3)^2=(\sqrt{5})^2$$
$$(x+1)^2+(y-3)^2=5$$

Be careful of the signs. It is easy to make a sign error in these steps.

6. Write the equation of the circle with center $(14,-5)$ and radius $\sqrt{7}$. Put your answer in standard form.

Copyright © 2013 Pearson Education, Inc.

Example	Student Practice
7. Write the equation of the circle $x^2 + 2x + y^2 + 6y + 6 = 0$ in standard form. Find the radius and center of the circle and sketch its graph.	**8.** Write the equation of the circle $x^2 + 4x - 2 + y^2 - 2y = 2$ in standard form. Find the radius and center of the circle and sketch its graph.

If we multiply out the terms in the standard form of the equation of a circle, we have the following.

$$(x-h)^2 + (y-k)^2 = r^2$$
$$\left(x^2 - 2hx + h^2\right) + \left(y^2 - 2ky + k^2\right) = r^2$$

Comparing this with the given equation, $\left(x^2 + 2x\right) + \left(y^2 + 6y\right) = -6,$ suggests we can complete the squares to put the equations in standard form.

$$x^2 + 2x + \underline{\quad} + y^2 + 6y + \underline{\quad} = -6$$
$$x^2 + 2x + 1 + y^2 + 6y + 9 = -6 + 1 + 9$$
$$x^2 + 2x + 1 + y^2 + 6y + 9 = 4$$
$$(x+1)^2 + (y+3)^2 = 4$$

Thus, the center is at $(-1, -3)$, and the radius is 2. The sketch of the circle is shown below.

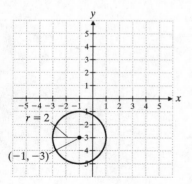

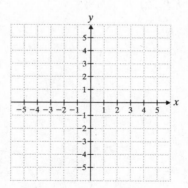

Copyright © 2013 Pearson Education, Inc.

Extra Practice

1. Find the distance between $(-0.5, 8.2)$ and $(3.5, 6.2)$.

2. Write the equation of the circle with center $\left(0, \dfrac{6}{5}\right)$ and radius $\sqrt{13}$. Put your answer in standard form.

3. Find the center and radius of the circle $x^2 + y^2 = 36$. Then sketch its graph.

4. Write the equation of the circle $x^2 + 10x + y^2 - 6y + 29 = 0$ in standard form. Find the radius and center of the circle.

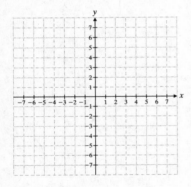

Concept Check

Explain how you would find the values of the unknown coordinate x if the distance between $(-6, 8)$ and $(x, 12)$ is 4.

Copyright © 2013 Pearson Education, Inc.

Name: _____   Date: _____

Instructor: _____   Section: _____

Chapter 9 The Conic Sections
9.2 The Parabola

Vocabulary
parabola • directrix • focus • axis of symmetry • vertex

1. The point at which the parabola crosses the axis of symmetry is the _____.

2. A(n) _____ is defined as the set of points in a plane that are a fixed distance from some fixed line and some fixed point that is not on the line.

3. The fixed point that a parabola is a fixed distance from is called a(n) _____.

Example	Student Practice
1. Graph $y = (x-2)^2$. Identify the vertex and the axis of symmetry.	**2.** Graph $y = (x+1)^2$. Identify the vertex and the axis of symmetry.

Make a table of values. Begin with $x = 2$ in the middle of the table of values because $(2-2)^2 = 0$. That is, when $x = 2$, $y = 0$. Then fill in the x- and y-values above and below $x = 2$.

x	4	3	2	1	0
y	4	1	0	1	4

Plot the points and draw the graph.

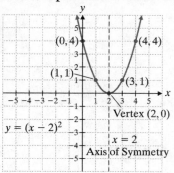

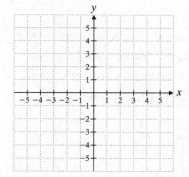

The vertex is $(2,0)$, and the axis of symmetry is the line $x = 2$.

Vocabulary Answers: 1. vertex 2. parabola 3. focus

Copyright © 2013 Pearson Education, Inc.

Example	Student Practice
3. Graph $y = -\frac{1}{2}(x+3)^2 - 1$.	**4.** Graph $y = -\frac{1}{4}(x+2)^2 - 4$.

Example (continued):

Rewrite the equation in standard form.

$$y = -\frac{1}{2}\left[x - (-3)\right]^2 + (-1)$$

Thus, $a = -\frac{1}{2}$, $h = -3$, and $k = -1$, so it is a vertical parabola. The parabola opens downward since $a < 0$. The vertex is $(-3, -1)$ and the axis of symmetry is the line $x = -3$. The y-intercept is $(0, -5.5)$. Plot a few points on either side of the axis of symmetry.

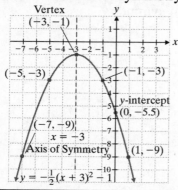

Student Practice 4:

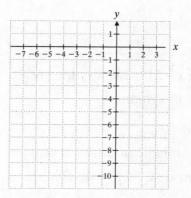

---

5. Graph $x = -2y^2$.

This is a horizontal parabola because the y term is squared. Make a table of values, but choose values for y instead of x. The graph is shown below.

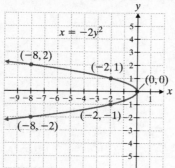

6. Graph $x = -3y^2$.

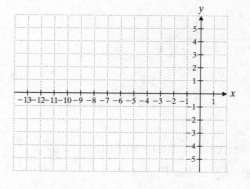

Copyright © 2013 Pearson Education, Inc.

Example

7. Place the equation $x = y^2 + 4y + 1$ in standard form. Then graph it.

$$x = y^2 + 4y + 1$$
$$= y^2 + 4y + \left(\frac{4}{2}\right)^2 - \left(\frac{4}{2}\right)^2 + 1$$
$$= (y+2)^2 - 3$$

Notice that $a = 1$, $k = -2$, and $h = -3$. Since a is positive, the parabola opens to the right. The vertex is $(-3, -2)$ and the axis of symmetry is $y = -2$. Let $y = 0$, to find the x-intercept, $(1, 0)$.

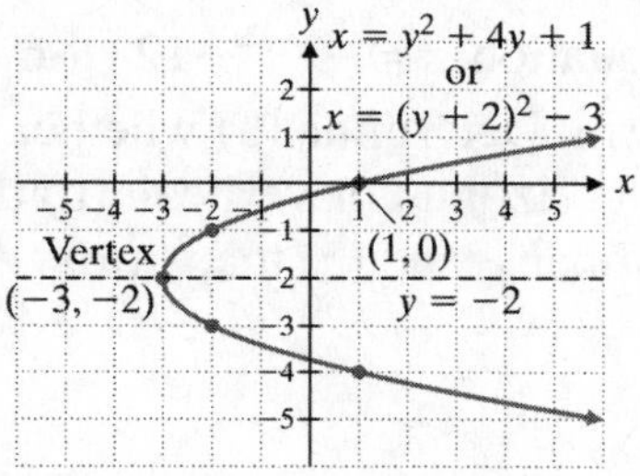

Student Practice

8. Place the equation $x = y^2 + 6y + 7$ in standard form. Then graph it.

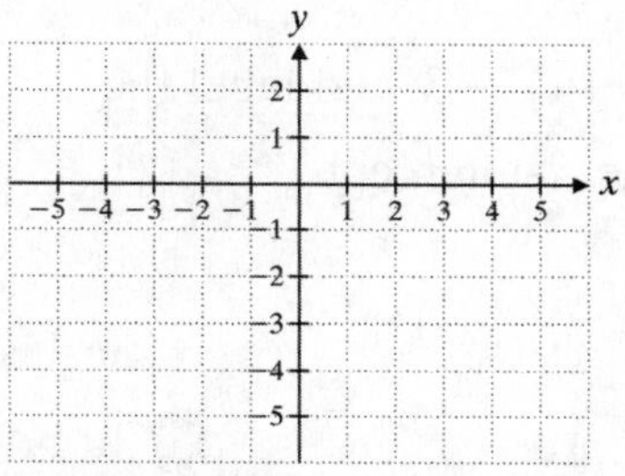

Example

9. Place the equation $y = 2x^2 - 4x - 1$ in standard form. Then graph it.

$$y = 2x^2 - 4x - 1$$
$$= 2\left[x^2 - 2x + (1)^2\right] - 2(1)^2 - 1$$
$$= 2(x-1)^2 - 3$$

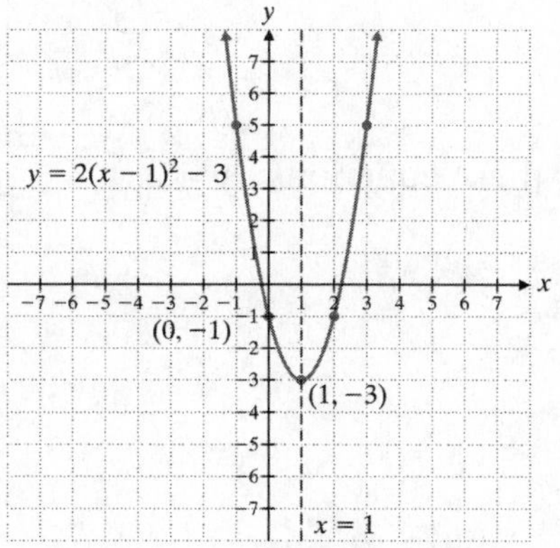

Student Practice

10. Place the equation $y = -2x^2 - 8x - 4$ in standard form. Then graph it.

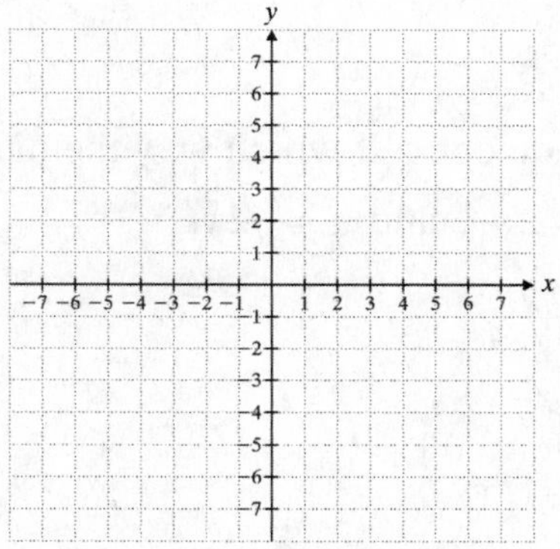

Copyright © 2013 Pearson Education, Inc.

1. Graph $y = 2(x+4)^2 - 3$ and label the vertex. Find the y-intercept.

2. Graph $y = -2\left(x - \dfrac{3}{2}\right)^2 + 3$ and label the vertex. Find the y-intercept.

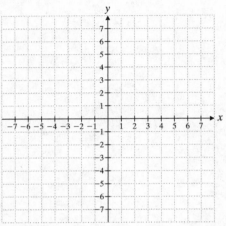

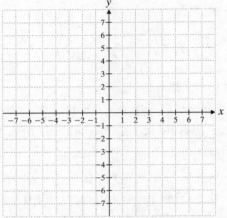

3. Graph $x = 3(y+1)^2 - 3$ and label the vertex. Find the x-intercept.

4. Place the equation $x = -3y^2 + 12y + 6$ in standard form. Determine (a) whether the parabola is horizontal or vertical, (b) the direction it opens, and (c) the vertex.

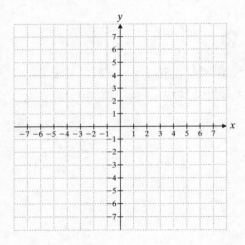

Concept Check

Explain how you can tell which way the parabola opens for these equations: $y = 2x^2$, $y = -2x^2$, $x = 2y^2$, and $x = -2y^2$.

Copyright © 2013 Pearson Education, Inc.

Name: _____   Date: _____

Instructor: _____   Section: _____

Chapter 9 The Conic Sections
9.3 The Ellipse

Vocabulary

ellipse   •   foci   •   vertices   •   center at the origin   •   center at (h, k)

1. The fixed points in an ellipse are called _____.

2. We define a(n) _____ as the set of points in a plane such that for each point in the set, the sum of its distances to two fixed points is constant.

Example	**Student Practice**
1. Graph $x^2 + 3y^2 = 12$. Label the intercepts.	**2.** Graph $9x^2 + 16y^2 = 144$. Label the intercepts.

Example

Rewrite the equation in standard form.

$$x^2 + 3y^2 = 12$$

$$\frac{x^2}{12} + \frac{3y^2}{12} = \frac{12}{12}$$

$$\frac{x^2}{12} + \frac{y^2}{4} = 1$$

Thus, we have the following:

$a^2 = 12$ so $a = 2\sqrt{3}$
$b^2 = 4$ so $b = 2$

The x-intercepts are $\left(-2\sqrt{3}, 0\right)$ and $\left(2\sqrt{3}, 0\right)$, and the y-intercepts are $(0, 2)$ and $(0, -2)$. We plot these points and draw the ellipse.

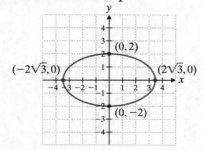

Student Practice

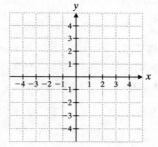

Vocabulary Answers: 1. foci 2. ellipse

Copyright © 2013 Pearson Education, Inc.

Example	Student Practice
3. Graph $\dfrac{(x-5)^2}{9} + \dfrac{(y-6)^2}{4} = 1$.	**4.** Graph $\dfrac{(x-2)^2}{16} + \dfrac{(y-3)^2}{4} = 1$.

Notice that this ellipse has the form

$$\frac{(x-h)^2}{a^2} + \frac{(y-k)^2}{b^2} = 1.$$

The center is (h,k). Note that a and b are not the x-intercepts and y-intercepts now. You'll see that a is the horizontal distance from the center of the ellipse to a point on the ellipse. Similarly, b is the vertical distance. Hence, when the center of the ellipse is not at the origin, the ellipse may not cross either axis.

The center of the ellipse is $(5,6)$, $a = 3$, and $b = 2$. Therefore, we begin at $(5,6)$. We plot points 3 units to the left, 3 units to the right, 2 units up, and 2 units down from $(5,6)$. The points we plot are the vertices of the ellipse.

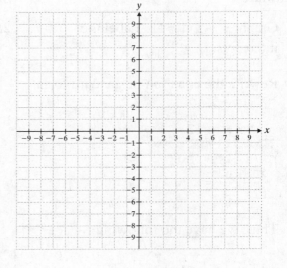

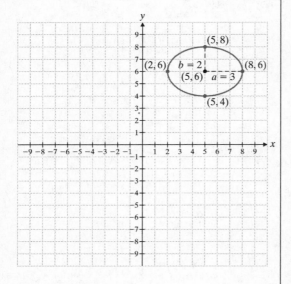

Copyright © 2013 Pearson Education, Inc.

Extra Practice

1. Graph $\dfrac{x^2}{4} + \dfrac{y^2}{36} = 1$. Label the intercepts.

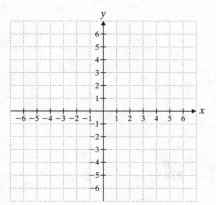

2. Graph $x^2 + 4y^2 = 16$. Label the intercepts.

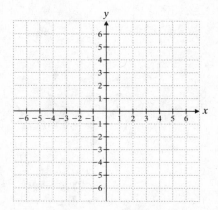

3. Graph $\dfrac{(x+2)^2}{16} + \dfrac{(y-3)^2}{9} = 1$. Label the center.

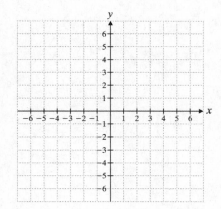

4. Graph $\dfrac{(x-2)^2}{9} + \dfrac{y^2}{16} = 1$. Label the center.

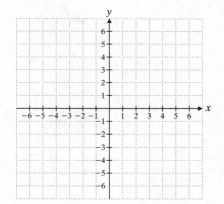

Concept Check

Explain how you would find the four vertices of the ellipse $12x^2 + y^2 - 36 = 0$.

Copyright © 2013 Pearson Education, Inc.

Copyright © 2013 Pearson Education, Inc.

Name: _____   Date: _____
Instructor: _____   Section: _____

Chapter 9 The Conic Sections
9.4 The Hyperbola

Vocabulary

hyperbola • foci • axis • vertices • asymptotes • fundamental rectangle

1. The points where the hyperbola intersects its axis are called the _____.

2. We define a(n) _____ as the set of points in a plane such that for each point in the set, the absolute value of the difference of its distances to two fixed points is constant.

Example	Student Practice
1. Graph $\dfrac{x^2}{25} - \dfrac{y^2}{16} = 1$.	**2.** Graph $\dfrac{x^2}{4} - \dfrac{y^2}{9} = 1$.

1. Graph $\dfrac{x^2}{25} - \dfrac{y^2}{16} = 1$.

The equation has the form $\dfrac{x^2}{a^2} - \dfrac{y^2}{b^2} = 1$, so it is a horizontal hyperbola. $a^2 = 25$, so $a = 5$; $b^2 = 16$, so $b = 4$. Since the hyperbola is horizontal, it has vertices $(a, 0)$ and $(-a, 0)$ or $(5, 0)$ and $(-5, 0)$.

To draw the asymptotes, we construct a fundamental rectangle with corners at $(5, 4)$, $(5, -4)$, $(-5, 4)$, and $(-5, -4)$.

We draw extended diagonals of the rectangle as the asymptotes. We construct each branch of the curve so that it passes through a vertex and gets closer to the asymptotes as it moves away from the origin.

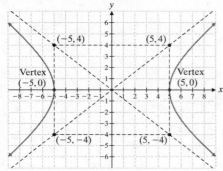

2. Graph $\dfrac{x^2}{4} - \dfrac{y^2}{9} = 1$.

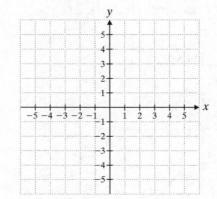

Vocabulary Answers: 1. vertices 2. hyperbola

Copyright © 2013 Pearson Education, Inc.

Example	Student Practice
3. Graph $4y^2 - 7x^2 = 28$.	**4.** Graph $3y^2 - 2x^2 = 18$.

3. Graph $4y^2 - 7x^2 = 28$.

To find the vertices and asymptotes, we must rewrite the equation in standard form. Divide each term by 28.

$$4y^2 - 7x^2 = 28$$

$$\frac{4y^2}{28} - \frac{7x^2}{28} = \frac{28}{28}$$

$$\frac{y^2}{7} - \frac{x^2}{4} = 1$$

Thus, we have the standard form of a vertical hyperbola with center at the origin.

Here $b^2 = 7$, so $b = \sqrt{7}$; $a^2 = 4$, so $a = 2$.

The hyperbola has vertices at $\left(0, \sqrt{7}\right)$ and $\left(0, -\sqrt{7}\right)$. The fundamental rectangle has corners at $\left(2, \sqrt{7}\right)$, $\left(2, -\sqrt{7}\right)$, $\left(-2, \sqrt{7}\right)$, and $\left(-2, -\sqrt{7}\right)$.

To aid us in graphing, we measure the distance $\sqrt{7}$ as approximately 2.6.

4. Graph $3y^2 - 2x^2 = 18$.

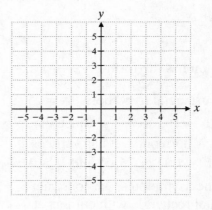

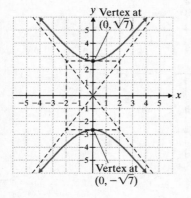

Copyright © 2013 Pearson Education, Inc.

Example	Student Practice

5. Graph $\dfrac{(x-4)^2}{9} - \dfrac{(y-5)^2}{4} = 1$.

6. Graph $\dfrac{(x-1)^2}{4} - \dfrac{(y-3)^2}{16} = 1$.

The center is at $(4,5)$, and the hyperbola is horizontal. We have $a = 3$ and $b = 2$, so the vertices are $(4 \pm 3, 5)$, or $(7,5)$ and $(1,5)$. We can sketch the hyperbola more readily if we draw a fundamental rectangle. Using $(4,5)$ as the center, we construct a rectangle $2a$ units wide and $2b$ units high. We then draw and extend the diagonals of the rectangle. The extended diagonals are the asymptotes for the branches of the hyperbola.

In this example, since $a = 3$ and $b = 2$, we draw a rectangle $2a = 6$ units wide and $2b = 4$ units high with a center at $(4,5)$. We draw extended diagonals through the rectangle. From the vertex at $(7,5)$, we draw a branch of the hyperbola opening to the right. From the vertex at $(1,5)$, we draw a branch of the hyperbola opening to the left. The graph of the hyperbola is shown.

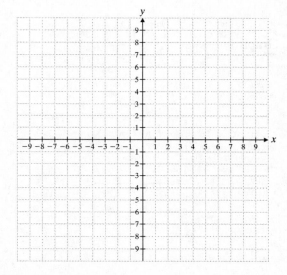

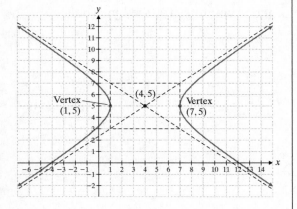

Copyright © 2013 Pearson Education, Inc.

Extra Practice

1. Graph $\dfrac{x^2}{4} - \dfrac{y^2}{36} = 1$.

2. Graph $y^2 - x^2 = 16$.

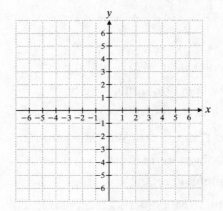

3. Graph $\dfrac{(x+2)^2}{4} - \dfrac{(y+1)^2}{9} = 1$.

4. Graph $\dfrac{(y+1)^2}{9} - \dfrac{(x-1)^2}{9} = 1$.

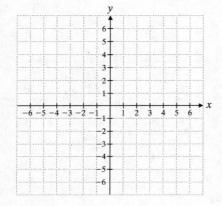

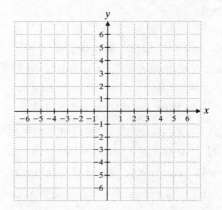

Concept Check
Explain how you would find the equation of one of the asymptotes for the hyperbola $49x^2 - 4y^2 = 196$.

Copyright © 2013 Pearson Education, Inc.

Chapter 9 The Conic Sections
9.5 Nonlinear Systems of Equations

Vocabulary
nonlinear equation   •   nonlinear system of equations

1. A _____ includes at least one nonlinear equation.

2. Any equation that is of second degree or higher is a _____.

Example	Student Practice
1. Solve the following nonlinear system and verify your answer with a sketch.	**2.** Solve the following nonlinear system and verify your answer with a sketch.

Example

1. Solve the following nonlinear system and verify your answer with a sketch.

$$x + y - 1 = 0 \qquad (1)$$
$$y - 1 = x^2 + 2x \quad (2)$$

We will use the substitution method. Solve for y in equation (1).

$$x + y - 1 = 0$$
$$y = -x + 1 \quad (3)$$

Substitute (3) into equation (2). Then solve the resulting quadratic equation.

$$y - 1 = x^2 + 2x$$
$$(-x + 1) - 1 = x^2 + 2x$$
$$-x + 1 - 1 = x^2 + 2x$$
$$0 = x^2 + 3x$$
$$0 = x(x + 3)$$
$$x = 0 \quad \text{or} \quad x = -3$$

Now substitute the values for x in the equation $y = -x + 1$.
Continued on the next page.

Student Practice

2. Solve the following nonlinear system and verify your answer with a sketch.

$$x + y - 5 = 0$$
$$y - 5 = x^2 + 10x$$

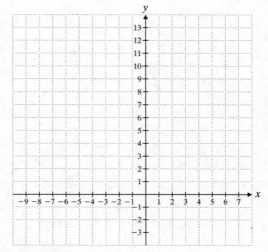

Vocabulary Answers: 1. nonlinear system of equations 2. nonlinear equation

Copyright © 2013 Pearson Education, Inc.

Example	Student Practice

For $x = -3$:

$y = -(-3)+1 = +3+1 = 4$

For $x = 0$:

$y = -(0)+1 = +1 = 1$

Thus, the solutions of the system are $(-3,4)$ and $(0,1)$.

Sketch the system. Equation (2) describes a parabola. Write it in the following form.

$y = x^2 + 2x + 1 = (x+1)^2$

This is a parabola opening upward with its vertex at $(-1,0)$. Equation (1) can be written as $y = -x+1$, which is a straight line with slope $= -1$ and y-intercept $(0,1)$.

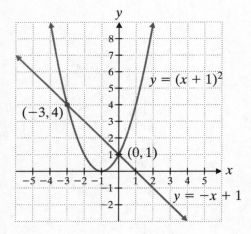

The sketch verifies that the two graphs intersect at $(-3,4)$ and $(0,1)$.

Copyright © 2013 Pearson Education, Inc.

Example	Student Practice
3. Solve the following nonlinear system and verify your answer with a sketch. $y - 2x = 0 \qquad (1)$ $\dfrac{x^2}{4} + \dfrac{y^2}{9} = 1 \qquad (2)$	**4.** Solve the following nonlinear system and verify your answer with a sketch. $y + 2x = 0$ $x^2 + \dfrac{y^2}{4} = 1$

Solving equation (1) for y yields

$y = 2x \quad (3)$.

Substitute (3) into equation (2).

$$\frac{x^2}{4} + \frac{(2x)^2}{9} = 1$$

$$36\left(\frac{x^2}{4}\right) + 36\left(\frac{(2x)^2}{9}\right) = 36(1)$$

$$9x^2 + 16x^2 = 36$$

$$x = \pm\sqrt{\frac{36}{25}}$$

$$x = \pm 1.2$$

For $x = +1.2$: $y = 2(1.2) = 2.4$.

For $x = -1.2$: $y = 2(-1.2) = -2.4$.

We recognize $\dfrac{x^2}{4} + \dfrac{y^2}{9} = 1$ as an ellipse

and $y = 2x$ as a straight line. The sketch shows that the points of intersection at $(1.2, 2.4)$ and $(-1.2, -2.4)$ are reasonable.

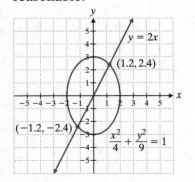

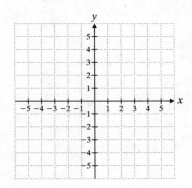

237

Copyright © 2013 Pearson Education, Inc.

Example	Student Practice
5. Solve the system. $$4x^2 + y^2 = 1 \quad (1)$$ $$x^2 + 4y^2 = 1 \quad (2)$$	**6.** Solve the system. $$9x^2 + y^2 = 1 \quad (1)$$ $$x^2 + 9y^2 = 1 \quad (2)$$

Since neither equation is linear, we will use the addition method. Multiply equation (1) by -4 and add to equation (2).

$$-16x^2 - 4y^2 = -4$$
$$\underline{x^2 + 4y^2 = 1}$$
$$-15x^2 = -3$$

$$x^2 = \frac{-3}{-15}$$

$$x^2 = \frac{1}{5}$$

$$x = \pm\sqrt{\frac{1}{5}}$$

If $x = +\sqrt{\dfrac{1}{5}}$, then $x^2 = \dfrac{1}{5}$. Substituting this value into equation (2) gives

$$\frac{1}{5} + 4y^2 = 1$$

$$4y^2 = \frac{4}{5}$$

$$y = \pm\sqrt{\frac{1}{5}}$$

Similarly, if $x = -\sqrt{\dfrac{1}{5}}$, then $y = \pm\sqrt{\dfrac{1}{5}}$. In this case, we have four solutions. If we rationalize each expression, the four

solutions are $\left(\dfrac{\sqrt{5}}{5}, \dfrac{\sqrt{5}}{5}\right)$, $\left(\dfrac{\sqrt{5}}{5}, -\dfrac{\sqrt{5}}{5}\right)$,

$\left(-\dfrac{\sqrt{5}}{5}, \dfrac{\sqrt{5}}{5}\right)$, and $\left(-\dfrac{\sqrt{5}}{5}, -\dfrac{\sqrt{5}}{5}\right)$.

Copyright © 2013 Pearson Education, Inc.

Extra Practice

1. Solve the following nonlinear system and verify your answer with a sketch.

$$y = x^2 - 4$$
$$y = x + 2$$

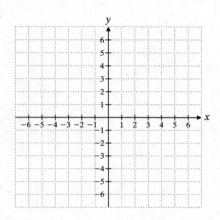

2. Solve the following nonlinear system by the substitution method.

$$x^2 - 49y^2 - 25 = 0$$
$$x + 7y - 2 = 0$$

3. Solve the following nonlinear system and verify your answer with a sketch.

$$x^2 + y^2 = 25$$
$$20x^2 - 5y^2 = 100$$

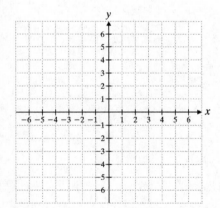

4. Solve the following nonlinear system by the addition method.

$$4x^2 = 3y^2 + 24$$
$$2(x^2 - 15) = -3y^2$$

Concept Check

Explain how you would solve the following system.

$$y^2 + 2x^2 = 18$$
$$xy = 4$$

Copyright © 2013 Pearson Education, Inc.

Copyright © 2013 Pearson Education, Inc.

MATH COACH

Mastering the skills you need to do well on the test.

Watch the **MATH COACH** videos in MyMathLab®or on You[Tube]™ while you work the problems below. These helpful hints will help you avoid making common errors on test problems.

Rewriting and Graphing the Equation of a Parabola—

Problem 2 Rewrite the equation in standard form. Find the center or vertex, plot at least one other point, identify the conic, and sketch the curve.

$$y^2 - 6y - x + 13 = 0$$

> **Helpful Hint:** If the equation of the parabola has a y^2 term, then it can be written in standard form $x = a(y-k)^2 + h$. The vertex is (h,k), and the graph is a horizontal parabola.

Did you first rewrite the equation as $x = y^2 - 6y + 13$?
Yes ____ No ____
Next, did you complete the square to obtain

$x = (y-3)^2 + 4$? Yes ____ No ____

If you answered No to these questions, remember that when completing the square, we take half of −6 and square it, which results in 9. We must add both a positive and a negative 9 to the right side of the equation. Go back and try to complete these steps again.

Did you see that since $a = 1$ and $a > 0$, the parabola opens to the right? Yes ____ No ____

Did you identify the vertex as $(4,3)$?
Yes ____ No ____

If you answered No to these questions, review the helpful hint and other information about horizontal parabolas. Stop and perform this step again.

In your final answer, be sure to give the equation in standard form, list the vertex, identify the conic, and graph the equation.

If you answered Problem 2 incorrectly, go back and rework the problem using these suggestions.

Identifying and Graphing a Hyperbola with a Center at the Origin—Problem 5

Identify and graph each conic section. Label the center and/or vertex as appropriate. $\dfrac{x^2}{10} - \dfrac{y^2}{9} = 1$

> **Helpful Hint:** The standard form of a hyperbola with the center at the origin and vertices $(-a,0)$ and $(a,0)$ has the equation $\dfrac{x^2}{a^2} - \dfrac{y^2}{b^2} = 1$.

Did you realize that this equation represents a horizontal hyperbola with vertices $\left(-\sqrt{10},0\right)$ and $\left(\sqrt{10},0\right)$?

Yes ____ No ____
If you answered No, review the rules for the standard form of an equation of a hyperbola with its center at the origin and read the Helpful Hint. Then remember that since $a^2 = 10$ and $a > 0$, $a = \sqrt{10}$. Likewise, if $b^2 = 9$ and $b > 0$, then $b = 3$.
Did you obtain a fundamental rectangle with corners at $\left(\sqrt{10},3\right)$, $\left(\sqrt{10},-3\right)$, $\left(-\sqrt{10},3\right)$, and $\left(-\sqrt{10},-3\right)$?
Yes ____ No ____

If you answered No, note that drawing a fundamental rectangle with corners at (a,b), $(a,-b)$, $(-a,b)$, and $(-a,-b)$ can help in creating the graph. The extended diagonals of the rectangle become asymptotes.

In your final answer, remember to identify the conic section, create its graph, and label the center and the vertices.

Now go back and rework the problem using these suggestions.

Copyright © 2013 Pearson Education, Inc.

Identifying and Graphing an Ellipse With a Center Not at the Origin—Problem 7

Identify and graph each conic section. Label the center and/or vertex as appropriate. $\dfrac{(x+2)^2}{16}+\dfrac{(y-5)^2}{4}=1$

> **Helpful Hint:** When an ellipse has a center that is not at the origin, the center has the coordinates (h,k) and
>
> the equation in standard form is $\dfrac{(x-h)^2}{a^2}+\dfrac{(y-k)^2}{b^2}=1$, where both a and b are greater than zero.

Did you determine that the center of the ellipse is at $(-2,5)$ with $a=4$ and $b=2$?
Yes _____ No _____

If you answered No, remember that the standard form of the equation involves $x-h$ and $y-k$, so you must be careful in determining the signs of h and k.

Did you start at the center and find points a units to the left, a units to the right, b units up, and b units down to plot the points $(-6,5)$, $(2,5)$, $(-2,7)$, and $(-2,3)$?
Yes _____ No _____

If you answered No, remember that to find these four points you need to find the following: $(h-a,k)$, $(h+a,k)$, $(h,k+b)$, and $(h,k-b)$.

Plot all four points and the center and label these on your graph. Then use the four points to make a sketch of the ellipse. Remember to identify the conic as an ellipse in your final answer.

If you answered Problem 7 incorrectly, go back and rework the problem using these suggestions.

Solving a System of Nonlinear Equations—Problem 16

Solve. $x^2+2y^2=15$
$x^2-y^2=6$

> **Helpful Hint:** When two equations in a system have the form $ax^2+by^2=c$, where a, b, and c are real numbers, then it may be easiest to solve the system by the addition method.

If you multiplied the second equation by 2 and added the result to the first equation, do you get $3x^2=27$?
Yes _____ No _____
Can you solve this equation for x to get $x=3$ and $x=-3$?
Yes _____ No _____

If you answered No to these questions, remember that when you add the two equations together, the y^2 term adds to 0.

If you substitute $x=3$ into the first equation, do you get the equation $9+2y^2=15$? Yes _____ No _____

Can you solve this equation for y to get $y=\sqrt{3}$ and $y=-\sqrt{3}$? Yes _____ No _____

If you answered No to these questions, try substituting $x=3$ into the equation again and be careful to avoid calculation errors. Remember that you must also perform this same step with $x=-3$. Since x is squared, your results for y should be the same.

Your final answer should have a total of four possible ordered pair solutions to this system.

Now go back and rework this problem using these suggestions.

Copyright © 2013 Pearson Education, Inc.

Name: _____   Date: _____

Instructor: _____   Section: _____

Chapter 10 Additional Properties of Functions
10.1 Function Notation

Vocabulary
function notation   •   free-fall   •   radius   •   surface area

1. The surface area of a sphere is a function of _____.

2. The approximate distance an object in _____ travels when there is no initial downward velocity is given by the distance function $d(t)=16t^2$.

Example	Student Practice
1. If $g(x)=5-3x$, find the following.	**2.** If $h(x)=6x-7$, find the following.
(a) $g(a)$	**(a)** $h(b)$
$g(a)=5-3a$	
(b) $g(a+3)$	
$g(a+3)=5-3(a+3)=5-3a-9$ $=-4-3a$	**(b)** $h(b+6)$
(c) $g(a)+g(3)$	
This requires us to find each addend separately, then add them together.	
$g(a)=5-3a$ $g(3)=5-3(3)=5-9=-4$	**(c)** $h(b)+h(6)$
$g(a)+g(3)=(5-3a)+(-4)$ $=5-3a-4$ $=1-3a$	
Notice that $g(a+3) \neq g(a)+g(3)$.	

Vocabulary Answers: 1. radius 2. free-fall

Copyright © 2013 Pearson Education, Inc.

Example	Student Practice

3. If $r(x) = \dfrac{4}{x+2}$, find $r(a+3) - r(a)$.
Express this result as one fraction.

$$r(a+3) - r(a) = \frac{4}{a+3+2} - \frac{4}{a+2}$$

$$= \frac{4}{a+5} - \frac{4}{a+2}$$

To express this as one fraction, we note that the LCD $= (a+5)(a+2)$.

$$r(a+3) - r(a)$$

$$= \frac{4(a+2)}{(a+5)(a+2)} - \frac{4(a+5)}{(a+2)(a+5)}$$

$$= \frac{4a+8}{(a+5)(a+2)} - \frac{4a+20}{(a+2)(a+5)}$$

$$= \frac{4a-4a+8-20}{(a+5)(a+2)} = \frac{-12}{(a+5)(a+2)}$$

4. If $k(x) = \dfrac{5}{x+6}$, find $k(a+2) - k(a)$.
Express this result as one fraction.

---

5. Let $f(x) = 3x - 7$.

Find $\dfrac{f(x+h) - f(x)}{h}$.

First find $f(x+h)$, then subtract $f(x)$.

$$f(x+h) = 3(x+h) - 7 = 3x + 3h - 7$$

$$f(x+h) - f(x)$$
$$= (3x + 3h - 7) - (3x - 7)$$
$$= 3x + 3h - 7 - 3x + 7$$
$$= 3h$$

Therefore, $\dfrac{f(x+h) - f(x)}{h} = \dfrac{3h}{h} = 3$.

6. Let $f(x) = 5x + 8$.

Find $\dfrac{f(x+h) - f(x)}{h}$.

Copyright © 2013 Pearson Education, Inc.

Example	Student Practice
7. The surface area of a sphere is given by $S = 4\pi r^2$ where r is the radius. If we use $\pi = 3.14$ as an approximation, this becomes $S = 4(3.14)r^2$, or $S = 12.56r^2$.	**8.** The surface area of a sphere is given by $S = 4\pi r^2$ where r is the radius. If we use $\pi = 3.14$ as an approximation, this becomes $S = 4(3.14)r^2$, or $S = 12.56r^2$.

(a) Find the surface area of a sphere with a radius of 3 centimeters.

Write surface area as a function of r and solve for $S(r)$ when $r = 3$.

$$S(3) = 12.56(3)^2 = 113.04 \text{ cm}^2$$

(a) Find the surface area of a sphere with a radius of 6 centimeters.

(b) Suppose that an error is made and the radius is calculated to be $(3+e)$ centimeters. Find an expression for the surface area as a function of the error e.

$$S(e) = 12.56(3+e)^2$$
$$= 113.04 + 75.36e + 12.56e^2$$

(b) Suppose that an error is made and the radius is calculated to be $(6+e)$ centimeters. Find an expression for the surface area as a function of the error e.

(c) Evaluate the surface area for $r = (3+e)$ cm when $e = 0.2$. Round your answer to the nearest hundredth of a cm. What is the difference in the surface area due to the error in measurement?

An error in measurement was made, so use the function found in part **(b)**.

$$S(0.2)$$
$$= 113.04 + 75.36(0.2) + 12.56(0.2)^2$$
$$= 128.6144$$
$$\approx 128.61 \text{ cm}^2$$

If the radius of 3 cm was incorrectly calculated as 3.2 cm, the surface area would be too large by approximately $128.61 - 113.04 = 15.57 \text{ cm}^2$.

(c) Evaluate the surface area for $r = (6+e)$ cm when $e = 0.3$. Round your answer to the nearest hundredth of a cm. What is the difference in the surface area due to the error in measurement?

Copyright © 2013 Pearson Education, Inc.

Extra Practice

1. If $g(x) = 5x^2 - 8x + 9$, find $g(a+1)$.   **2.** If $h(x) = \sqrt{x+2}$, find $h(4a^2 + 6)$.

3. If $s(x) = \dfrac{5}{x+2}$, find $s\left(-\dfrac{4}{3}\right) + s(-3)$.   **4.** Find $\dfrac{f(x+h) - f(x)}{h}$ for $f(x) = 4x^2$.

Concept Check

Explain how you would find $k(2a-1)$ if $k(x) = \sqrt{3x+1}$.

Copyright © 2013 Pearson Education, Inc.

Chapter 10 Additional Properties of Functions
10.2 General Graphing Procedures for Functions

Vocabulary

function   •   vertical line test   •   vertical shift   •   horizontal shift

1. $f(x) + k$ represents the graph of $f(x)$ with a _____ upwards of k units.

2. A _____ must have no ordered pairs that have the same first coordinates and different second coordinates.

3. The _____ states that if any vertical line intersects the graph of a relation more than once, the relation is not a function.

4. $f(x+h)$ represents the graph of $f(x)$ with a _____ to the left of h units.

Example	Student Practice
1. Determine whether the following is the graph of a function.	**2.** Determine whether the following is the graph of a function.

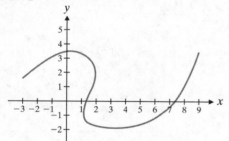

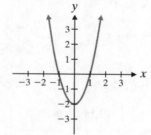

A vertical line intersects the graph more than once, so by the vertical line test, this relation is not a function.

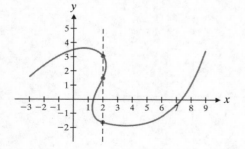

Vocabulary Answers: 1. vertical shift 2. function 3. vertical line test 4. horizontal shift

Copyright © 2013 Pearson Education, Inc.

Example	Student Practice

Example

3. Graph the functions on one coordinate plane. $f(x) = x^2$ and $h(x) = x^2 + 2$

First we make a table of values for $f(x)$ and for $h(x)$.

x	$f(x) = x^2$
−2	4
−1	1
0	0
1	1
2	4

x	$h(x) = x^2 + 2$
−2	6
−1	3
0	2
1	3
2	6

Graph on the same coordinate plane.

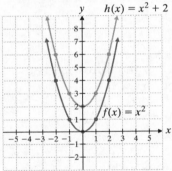

Notice that the graph of $h(x)$ is the graph of $f(x)$ moved 2 units upward.

5. Graph the functions on one coordinate plane. $f(x) = |x|$ and $p(x) = |x - 3|$

Graph each function on the same coordinate plane.

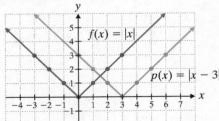

The graph of $p(x)$ is the graph of $f(x)$ shifted 3 units to the right.

Student Practice

4. Graph the functions on one coordinate plane. $f(x) = x^2$ and $h(x) = x^2 - 3$

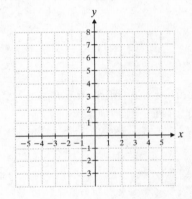

6. Graph the functions on one coordinate plane. $f(x) = |x|$ and $p(x) = |x + 3|$

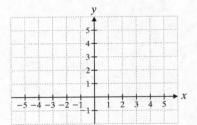

Copyright © 2013 Pearson Education, Inc.

Example	Student Practice
7. Graph the functions on one coordinate plane. $f(x) = x^3$ and $h(x) = (x-3)^3 - 2$	**8.** Graph the functions on one coordinate plane. $f(x) = x^3$ and $h(x) = (x+4)^3 - 2$

First we make a table of values for $f(x)$ and graph the function.

x	f(x)
−2	−8
−1	−1
0	0
1	1
2	8

Next we recognize that $h(x)$ will have a similar shape, but the curve will be shifted 3 units to the right and 2 units downward. We draw the graph of $h(x)$ using these shifts.

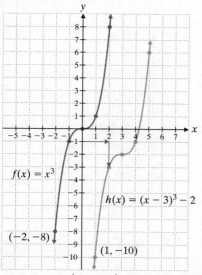

$f(x) = x^3$

$h(x) = (x-3)^3 - 2$

$(-2, -8)$

$(1, -10)$

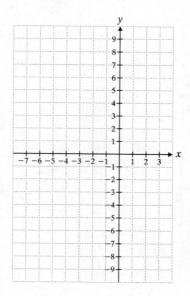

The point $(-2, -8)$ has been shifted 3 units to the right and 2 units down to the point $(-2+3, -8+(-2))$ or $(1, -10)$.

The point $(-1, -1)$ is a point on $f(x)$. Use the same reasoning to find the image of $(-1, -1)$ on the graph of $h(x)$. Verify by checking the graphs.

249

Copyright © 2013 Pearson Education, Inc.

Extra Practice

1. Determine whether the following is the graph of a function.

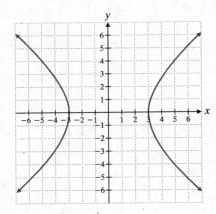

2. Graph the functions on one coordinate plane. $f(x) = x^2$ and $g(x) = (x+3)^2 - 2$

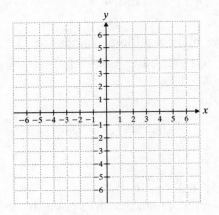

3. Graph the functions on one coordinate plane. $f(x) = x^3$ and $g(x) = (x+2)^3 + 2$

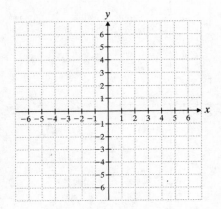

4. Graph the functions on one coordinate plane. $f(x) = \dfrac{3}{x}$ and $g(x) = \dfrac{3}{x-2}$

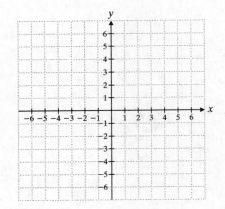

Concept Check
Explain how you can use a vertical line to determine whether a graph represents a function.

Copyright © 2013 Pearson Education, Inc.

Chapter 10 Additional Properties of Functions
10.3 Algebraic Operations on Functions

Vocabulary

sum   •   difference   •   product   •   quotient   •   composition

1. When finding the _____ of a function, we must be careful to avoid division by zero.

2. The _____ of the functions f and g, is defined as follows:
    $(fg)(x) = f(x) \cdot g(x)$.

3. The _____ of the functions f and g, denoted $f \circ g$, is defined as follows:
    $(f \circ g)(x) = f[g(x)]$.

4. The _____ of the functions f and g, is defined as follows:
    $(f + g)(x) = f(x) + g(x)$.

Example	Student Practice
1. Suppose that $f(x) = 3x^2 - 3x + 5$ and $g(x) = 5x - 2$. (a) Find $(f + g)(x)$. $\begin{aligned}(f+g)(x) &= f(x) + g(x) \\ &= (3x^2 - 3x + 5) + (5x - 2) \\ &= 3x^2 + 2x + 3 \end{aligned}$ (b) Evaluate $(f + g)(x)$ when $x = 3$. Write $(f + g)(3)$ and use the formula obtained in (a). $\begin{aligned}(f + g)(x) &= 3x^2 + 2x + 3 \\ (f + g)(3) &= 3(3)^2 + 2(3) + 3 \\ &= 27 + 6 + 3 = 36 \end{aligned}$	2. Suppose that $f(x) = 2x^2 + 4x - 7$ and $g(x) = 8x + 1$. (a) Find $(f + g)(x)$. (b) Evaluate $(f + g)(x)$ when $x = 4$.

Vocabulary Answers: 1. quotient 2. composition 3. product 4. sum

Copyright © 2013 Pearson Education, Inc.

Example	Student Practice
3. Given $f(x) = x^2 - 5x + 6$ and $g(x) = 2x - 1$, find the following.	**4.** Given $f(x) = x^2 - 3x + 8$ and $g(x) = 4x - 3$, find the following.
(a) $(fg)(x)$	**(a)** $(fg)(x)$

$$(fg)(x) = f(x) \cdot g(x)$$
$$= (x^2 - 5x + 6)(2x - 1)$$
$$= 2x^3 - 11x^2 + 17x - 6$$

(b) $(fg)(-4)$ — **(b)** $(fg)(-2)$

Use the formula obtained in **(a)**.

$$(fg)(x) = 2x^3 - 11x^2 + 17x - 6$$
$$(fg)(-4)$$
$$= 2(-4)^3 - 11(-4)^2 + 17(-4) - 6$$
$$= -128 - 176 - 68 - 6 = -378$$

---

5. Given $f(x) = 3x + 1$, $g(x) = 2x - 1$, and $h(x) = 9x^2 + 6x + 1$, find the following.

6. Given $f(x) = 2x + 3$, $g(x) = x - 4$, and $h(x) = 4x^2 + 12x + 9$, find the following.

(a) $\left(\dfrac{f}{g}\right)(x)$ — **(a)** $\left(\dfrac{f}{g}\right)(x)$

$$\left(\frac{f}{g}\right)(x) = \frac{3x + 1}{2x - 1}$$

The denominator of the quotient can never be zero. Since $2x - 1 \neq 0$, we know that $x \neq \dfrac{1}{2}$.

(b) $\left(\dfrac{f}{h}\right)(x)$ — **(b)** $\left(\dfrac{f}{h}\right)(x)$

$$\left(\frac{f}{h}\right)(x) = \frac{3x + 1}{9x^2 + 6x + 1}$$

$$= \frac{3x + 1}{(3x + 1)(3x + 1)} = \frac{1}{3x + 1}$$

Since $3x + 1 \neq 0$, we know $x \neq -\dfrac{1}{3}$.

Copyright © 2013 Pearson Education, Inc.

Example	Student Practice
7. Given $f(x) = 3x - 2$ and $g(x) = 2x + 5$, find $f[g(x)]$.	**8.** Given $f(x) = 5x + 3$ and $g(x) = 3x - 4$, find $f[g(x)]$.

First substitute $g(x) = 2x + 5$. Then apply the formula for $f(x)$. Remove the parentheses and simplify.

$$f[g(x)] = f(2x + 5)$$
$$= 3(2x + 5) - 2$$
$$= 6x + 15 - 2$$
$$= 6x + 13$$

---

9. Given $f(x) = \sqrt{x - 4}$ and $g(x) = 3x + 1$, find the following.

(a) $f[g(x)]$

Substitute $g(x) = 3x + 1$. Then apply the formula for $f(x)$.

$$f[g(x)] = f(3x + 1)$$
$$= \sqrt{(3x + 1) - 4}$$
$$= \sqrt{3x + 1 - 4}$$
$$= \sqrt{3x - 3}$$

(b) $g[f(x)]$

$$g[f(x)] = g\left(\sqrt{x - 4}\right)$$
$$= 3\left(\sqrt{x - 4}\right) + 1$$
$$= 3\sqrt{x - 4} + 1$$

We note that $g[f(x)] \neq f[g(x)]$.

10. Given $f(x) = \sqrt{x + 6}$ and $g(x) = 4x - 1$, find the following.

(a) $f[g(x)]$

(b) $g[f(x)]$

Copyright © 2013 Pearson Education, Inc.

Example	Student Practice
11. Given $f(x) = 2x$ and $g(x) = \dfrac{1}{3x-4}$, $x \neq \dfrac{4}{3}$, find the following. **(a)** $(f \circ g)(x)$	**12.** Given $f(x) = 4x$ and $g(x) = \dfrac{1}{2x-5}$, $x \neq \dfrac{5}{2}$, find the following. **(a)** $(f \circ g)(x)$

$$(f \circ g)(x) = f\big[g(x)\big] = f\left(\frac{1}{3x-4}\right)$$

$$= 2\left(\frac{1}{3x-4}\right) = \frac{2}{3x-4}$$

(b) $(f \circ g)(2)$

(b) $(f \circ g)\left(\dfrac{1}{2}\right)$

$$(f \circ g)(2) = \frac{2}{3(2)-4} = \frac{2}{6-4} = \frac{2}{2} = 1$$

Extra Practice

1. Given $f(x) = 1.6x^3 - 2.7x$ and $g(x) = 4.6x^2 + 7.6$, find the following.

(a) $(f+g)(x)$
(b) $(f-g)(x)$
(c) $(f+g)(3)$
(d) $(f-g)(-2)$

2. Given $f(x) = x^2 - 8x + 16$ and $g(x) = x - 4$, find the following.

(a) $(fg)(x)$
(b) $\left(\dfrac{f}{g}\right)(x)$
(c) $(fg)(3)$
(d) $\left(\dfrac{f}{g}\right)(-3)$

3. Given $f(x) = x - 5$ and $g(x) = 6 - 2x$, find $f\big[g(x)\big]$.

4. Given $f(x) = \left|\dfrac{4}{3}x + 1\right|$ and $g(x) = -3x - 2$, find $f\big[g(x)\big]$.

Concept Check
If $f(x) = 5 - 2x^2$ and $g(x) = 3x^2 - 5x + 1$, explain how you would find $(f-g)(-4)$.

Copyright © 2013 Pearson Education, Inc.

Chapter 10 Additional Properties of Functions
10.4 Inverse of a Function

Vocabulary

inverse function f •   one-to-one function   •   horizontal line test

1. The _____ states that if any horizontal line intersects the graph of a function more than once, the function is not one-to-one.

2. We call a function f^{-1} that reverses the domain and range of a function f the _____.

3. A(n) _____ is a function in which no ordered pairs have the same second coordinate.

Example	Student Practice
1. Indicate whether the following functions are one-to-one.	**2.** Indicate whether the following functions are one-to-one.
(a) $M = \{(1,3),(2,7),(5,8),(6,12)\}$	**(a)** $L = \{(1,6),(4,3),(7,6),(9,16)\}$
M is a function because no ordered pairs have the same first coordinate. M is also a one-to-one function because no ordered pairs have the same second coordinate.	
(b) $P = \{(1,4),(2,9),(3,4),(4,18)\}$	**(b)** $N = \{(2,6),(3,4),(6,9),(7,11)\}$
P is a function, but it is not one-to-one because the ordered pairs $(1,4)$ and $(3,4)$ have the same second coordinate.	

Vocabulary Answers: 1. horizontal line test 2. inverse function f 3. one-to-one function

Copyright © 2013 Pearson Education, Inc.

Example	Student Practice
3. Determine whether the functions graphed are one-to-one functions.	**4.** Determine whether the functions graphed are one-to-one functions.

(a)

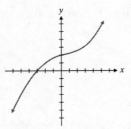

This graph represents a one-to-one function. Horizontal lines cross the graph at most once.

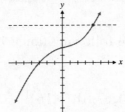

(a)

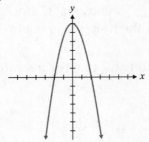

(b)

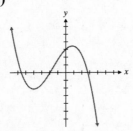

This graph does not represent a one-to-one function. A horizontal line exists that crosses the graph more than once.

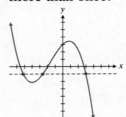

(b)

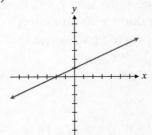

Copyright © 2013 Pearson Education, Inc.

Example	Student Practice
5. Determine the inverse function of $F = \{(6,1),(12,2),(13,5),(14,6)\}$. Since we have a list of ordered pairs, we interchange the coordinates of each ordered pair. The inverse function of F is as follows. $F^{-1} = \{(1,6),(2,12),(5,13),(6,14)\}$	**6.** Determine the inverse function of $Q = \{(2,4),(6,2),(9,7),(11,3)\}$.
7. Find the inverse of $f(x) = 7x - 4$. Replace $f(x)$ with y. $y = 7x - 4$ Interchange the variables x and y. $x = 7y - 4$ Solve for y in terms of x. $x = 7y - 4$ $x + 4 = 7y$ $\dfrac{x+4}{7} = y$ Replace y with $f^{-1}(x)$. $f^{-1}(x) = \dfrac{x+4}{7}$	**8.** Find the inverse of $f(x) = 5x + 8$.

Copyright © 2013 Pearson Education, Inc.

Example	Student Practice
9. Find the inverse function of $f(x) = \dfrac{9}{5}x + 32$, which converts Celsius temperature (x) into equivalent Fahrenheit temperature.	**10.** Find the inverse function of $f(x) = 150 + 3(x - 25)$, which gives the cost of a rental truck if the company charges a base rate of \$150 plus \$3 for every mile traveled over 25 miles. Here x is the total number of miles traveled in the rental truck.

$$y = \frac{9}{5}x + 32$$

$$x = \frac{9}{5}y + 32$$

$$5x = 9y + 160$$

$$\frac{5x - 160}{9} = y$$

$$f^{-1}(x) = \frac{5x - 160}{9}$$

Our inverse function $f^{-1}(x)$ will now convert Fahrenheit temperature to Celsius temperature.

---

11. If $f(x) = 3x - 2$, find $f^{-1}(x)$. Graph f and f^{-1} on the same set of axes. Draw the line $y = x$ as a dashed line for reference.

Following the procedure to find $f^{-1}(x)$ yields $f^{-1}(x) = \dfrac{x + 2}{3}$. Now we graph each line.

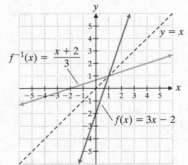

We see that the graphs of f and f^{-1} are symmetric about the line $y = x$.

12. If $f(x) = 4x + 1$, find $f^{-1}(x)$. Graph f and f^{-1} on the same set of axes. Draw the line $y = x$ as a dashed line for reference.

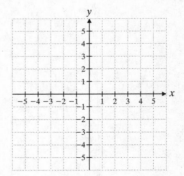

Copyright © 2013 Pearson Education, Inc.

Extra Practice

1. Indicate whether the following function is one-to-one.

$$B = \{(7,9),(9,7),(-7,-9),(-9,-7)\}$$

2. Find the inverse of $f(x) = x - 4$.

3. Find the inverse of $f(x) = \dfrac{5}{3x-4}$.

4. Find the inverse of $g(x) = -3x - 4$.
   Graph the function and its inverse on the same set of axes. Draw the line $y = x$ as a dashed line for reference.

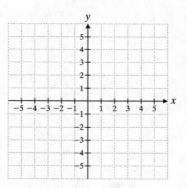

Concept Check

If $f(x) = \dfrac{x-5}{3}$, explain how you would find the inverse function.

Copyright © 2013 Pearson Education, Inc.

Copyright © 2013 Pearson Education, Inc.

MATH COACH

Mastering the skills you need to do well on the test.

Watch the **MATH COACH** videos in MyMathLab®or on You Tube while you work the problems below. These helpful hints will help you avoid making common errors on test problems.

Using Function Notation to Evaluate Expressions—

Problem 5 For the function $f(x) = 3x^2 - 2x + 4$, find $f(a+1)$.

> **Helpful Hint:** The key idea is to replace every x with the expression $a+1$ and then simplify the result. Use parentheses around the substitutions to avoid calculation errors.

Did you substitute $a+1$ into the function to get

$3(a+1)^2 - 2(a+1) + 4$?

Yes _____ No _____

Did you evaluate $(a+1)^2$ to get $a^2 + 2a + 1$?

Yes _____ No _____

If you answered No to these questions, remember to replace every x with $a+1$. Use parentheses to avoid calculation errors. Note that when you square a binomial, you must be sure to write down all the terms.

Next, did you simplify further to obtain

$3a^2 + 6a + 3 - 2a - 2 + 4$? Yes _____ No _____

If you answered No, remember to multiply all three terms of $a^2 + 2a + 1$ by 3. Multiply both terms of $a+1$ by -2.

In your final step, you can collect like terms to write your answer in simplest form.

If you answered Problem 5 incorrectly, go back and rework the problem using these suggestions.

Graphing a Function with a Horizontal and Vertical Shift—Problem 10 Graph each pair of functions on one coordinate plane. $f(x) = x^2$

$$g(x) = (x-1)^2 + 3$$

Can you determine that the graph of $f(x)$ passes through the points $(0,0)$, $(1,1)$, $(-1,1)$, $(2,4)$, and $(-2,4)$?

Yes _____ No _____

If you answered No, try building a table of values in which you replace x with 0, 1, -1, 2, and -2 and find the corresponding values of $f(x)$. Plot those points and connect the points with a curve to form the graph of $f(x) = x^2$.

Do you see that the function $g(x) = (x-1)^2 + 3$ has the values of $h = 1$ and $k = 3$ when you apply the Helpful Hint?

Yes _____ No _____

> **Helpful Hint:** First graph $f(x)$. The graph of $f(x-h) + k$ is the graph of $f(x)$ shifted h units to the right and k units upward (assuming that $h > 0$ and $k > 0$).

Did you find that the graph of $g(x)$ is the graph of $f(x)$ shifted one unit to the right and 3 units up? Yes _____ No _____

If you answered No to these questions, reread the Helpful Hint carefully. Notice that the values of h and k are both greater than zero.

Finish by graphing $g(x)$ on the same coordinate plane as your graph of $f(x)$.

Now go back and rework the problem using these suggestions.

Copyright © 2013 Pearson Education, Inc.

Finding the Composition of Two Functions—Problem 14(a)

If $f(x) = \frac{1}{2}x - 3$ and $g(x) = 4x + 5$, find $(f \circ g)(x)$.

Helpful Hint: First rewrite $(f \circ g)(x)$ as $f[g(x)]$. Most students find this expression more logical. Then substitute $g(x)$ for the value of x in $f(x)$.

First, did you rewrite the problem as

$$f[g(x)] = \frac{1}{2}(4x + 5) - 3?$$

Yes _____ No _____

If you answered No, substitute the expression for $g(x)$ for the value of x in the expression for $f(x)$.

Next did you simplify the resulting expression to

$$f[g(x)] = 2x + \frac{5}{2} - 3?$$

Yes _____ No _____

If you answered No, remember that

$\frac{1}{2}(4x) = 2x$ and $\frac{1}{2}(5) = \frac{5}{2}$. As your final

step, combine like terms to write your expression in simplest form.

If you answered Problem 14(a) incorrectly, go back and rework the problem using these suggestions.

Finding and Graphing the Inverse of a Function—Problem 18   Given $f(x) = -3x + 2$, find f^{-1}.

Graph f and its inverse f^{-1} on one coordinate plane. Graph $y = x$ as a dashed line for reference.

Helpful Hint: Use the following four steps to find the inverse of a function:
1. Replace $f(x)$ with y.
2. Interchange x and y.
3. Solve for y in terms of x.
4. Replace y with $f^{-1}(x)$.

Did you substitute y for $f(x)$ to get $y = -3x + 2$ and then interchange x and y to get $x = -3y + 2$?

Yes _____ No _____

If you answered No, review the first two steps in the Helpful Hint and perform these steps again.

Did you solve the equation for y to get $y = -\frac{1}{3}x + \frac{2}{3}$?

Yes _____ No _____

If you answered No, remember to add $3y$ to each side. Next add $-x$ to each side and then divide each side of the equation by 3. As your last step in finding the inverse, replace y with $f^{-1}(x)$.

Remember to graph $f(x)$ and $f^{-1}(x)$ on the same coordinate plane. Add the graph of $y = x$ as a dashed line for reference.

Now go back and rework the problem using these suggestions.

Copyright © 2013 Pearson Education, Inc.

Name: _____   Date: _____

Instructor: _____   Section: _____

Chapter 11 Logarithmic and Exponential Functions
11.1 The Exponential Function

Vocabulary
exponential function • e • asymptote • property of exponential equations
compound interest • radioactive decay • base

1. The _____ says that if $b^x = b^y$, then $x = y$ for $b > 0$ and $b \neq 1$.

2. The function $f(x) = b^x$, where $b > 0$, $b \neq 1$, and x is a real number, is called a(n) _____.

3. For every exponential function, the x-axis is a(n) _____.

Example	Student Practice
1. Graph $f(x) = 2^x$.	**2.** Graph $f(x) = 5^x$.

Make a table of values for x and $f(x)$.

$$f(-1) = 2^{-1} = \frac{1}{2},\ f(0) = 2^0 = 1,\ f(1) = 2^1 = 2$$

Continue to evaluate $f(x)$ at different integers to make a table of values for x and $f(x)$. Using these points, draw the graph.

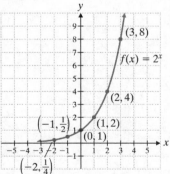

Notice how the curve comes very close to the x-axis but never touches it. The x-axis is an asymptote for every exponential function.

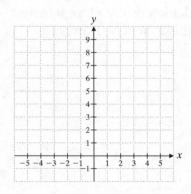

Vocabulary Answers: 1. property of exponential equations 2. exponential function 3. asymptote

Copyright © 2013 Pearson Education, Inc.

Example	Student Practice
3. Graph $f(x) = \left(\dfrac{1}{2}\right)^x$.	**4.** Graph $f(x) = \left(\dfrac{2}{3}\right)^x$.

3. Graph $f(x) = \left(\dfrac{1}{2}\right)^x$.

Rewrite the function as follows.

$$f(x) = \left(\dfrac{1}{2}\right)^x = \left(2^{-1}\right)^x = 2^{-x}$$

Evaluate the function for a few values of x. Using these points, draw the graph.

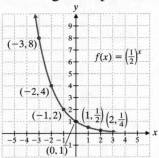

Note that as x increases, $f(x)$ decreases.

4. Graph $f(x) = \left(\dfrac{2}{3}\right)^x$.

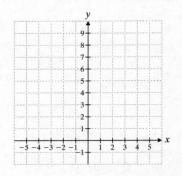

5. Graph $f(x) = e^x$.

The letter e is like the number π. It is an irrational number. If you do not have a scientific calculator, approximate the value for e to use the number in calculations, $e \approx 2.7183$. Otherwise, use a scientific calculator to produce a table of values for this function. Then, draw the graph.

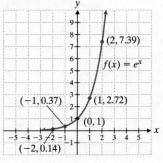

6. Graph $f(x) = e^{1+x}$.

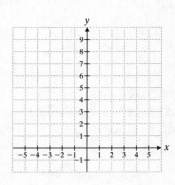

Copyright © 2013 Pearson Education, Inc.

Example	Student Practice
7. Solve. $2^x = \dfrac{1}{16}$	**8.** Solve. $3^x = 27$

7. To use the property of exponential equations, we must have the same base on both sides of the equation. Rewrite 16 as 2^4.

$2^x = \dfrac{1}{16}$

$2^x = \dfrac{1}{2^4}$

$2^x = 2^{-4}$

Recall the property of exponential equations, which states that if $b^x = b^y$, then $x = y$ for $b > 0$ and $b \neq 1$. Since $2^x = 2^{-4}$, $x = -4$.

---

9. If we invest \$8000 in a fund that pays 15% annual interest compounded monthly, how much will we have in 6 years?

In this situation, $P = 8000$, $r = 15\% = 0.15$, $t = 6$, and $n = 12$ because interest is compounded monthly, or 12 times a year.

$A = 8000\left(1 + \dfrac{0.15}{12}\right)^{(12)(6)}$

$\quad = 8000(1 + 0.125)^{72}$

$\quad = 8000(2.445920268)$

$\quad \approx 19,567.36$

10. If you invest \$5500 in a fund that pays 18% annual interest compounded quarterly, how much will you have in 10 years?

Copyright © 2013 Pearson Education, Inc.

Example	Student Practice
11. The radioactive decay of the element americium 241 can be described by the equation $A = Ce^{-0.001608t}$, where C is the original amount of the element in the sample, A is the amount of the element remaining after t years, and $k = -0.0016008$, the decay constant for americium 241. If 10 milligrams (mg) of americium 241 is sealed in a laboratory container today, how much will theoretically be present in 2000 years? Round your answer to the nearest hundredth.	**12.** The radioactive decay of the element cobalt 60 can be described by the equation $A = Ce^{-0.131527t}$, where C is the original amount of the element in the sample, A is the amount of the element remaining after t years, and $k \approx -0.131527$, the decay constant for cobalt 60. If 5 milligrams (mg) of cobalt 60 is sealed in a laboratory container today, how much will theoretically be present in 10 years? Round your answer to the nearest hundredth.

Substitute $C = 10$ and $t = 2000$ into the equation and simplify using a calculator.

$$A = Ce^{-0.0016008t} = 10e^{-0.0016008(2000)}$$
$$\approx 10(0.040697)$$
$$\approx 0.41 \text{ mg}$$

Extra Practice

1. Graph $f(x) = 3^{-x}$.

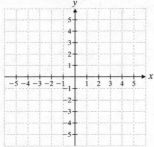

2. Graph $f(x) = 2^{x+4}$.

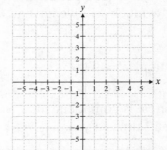

3. Solve for x. $5^{x+3} = 25$

4. Anton is investing $4000 at an annual rate of 4.6% compounded annually. How much money will Anton have after 5 years? Round your answer to the nearest cent.

Concept Check

Explain how you would solve $4^{-x} = \dfrac{1}{64}$.

Copyright © 2013 Pearson Education, Inc.

Chapter 11 Logarithmic and Exponential Functions
11.2 The Logarithmic Function

Vocabulary
logarithmic function • logarithm • base • power • inverse

1. The logarithmic function $y = \log_b x$ is the _____ of the exponential function $x = b^y$.

2. A(n) _____ is an exponent.

3. In the function $y = \log_b x$, b is called the _____.

Example	Student Practice
1. Write in logarithmic form. $81 = 3^4$ Use the fact that $x = b^y$ is equivalent to $\log_b x = y$. Here, $x = 81$, $b = 3$, and $y = 4$. So the logarithmic form of $81 = 3^4$ is $4 = \log_3 81$.	**2.** Write in logarithmic form. $\dfrac{1}{216} = 6^{-3}$
3. Write in exponential form. $-4 = \log_{10}\left(\dfrac{1}{10,000}\right)$ Use the fact that $x = b^y$ is equivalent to $\log_b x = y$. Here, $y = -4$, $b = 10$, and $x = \dfrac{1}{10,000}$. So the exponential form of $-4 = \log_{10}\left(\dfrac{1}{10,000}\right)$ is $\dfrac{1}{10,000} = 10^{-4}$.	**4.** Write in exponential form. $5 = \log_3 243$

Vocabulary Answers: 1. inverse 2. logarithm 3. base

Copyright © 2013 Pearson Education, Inc.

Example	Student Practice
5. Solve for the variable.	**6.** Solve for the variable
(a) $\log_5 x = -3$	**(a)** $\log_2 x = -8$
Convert the logarithmic equation to an equivalent exponential equation and solve for x.	
$5^{-3} = x$	
$\dfrac{1}{5^3} = x$	
$\dfrac{1}{125} = x$	
(b) $\log_a 16 = 4$	**(b)** $\log_b 64 = 3$
$a^4 = 16$	
$a^4 = 2^4$	
$a = 2$	
7. Evaluate. $\log_3 81$	**8.** Evaluate. $\log_2 64$
This is asking, "To what power must we raise 3 to get 81?" Write an equivalent exponential expression using x as the unknown power.	
$\log_3 81 = x$	
$81 = 3^x$	
Write 81 as 3^4 and solve for x.	
$3^4 = 3^x$	
$x = 4$	
Thus, $\log_3 81 = 4$.	

268

Copyright © 2013 Pearson Education, Inc.

Example	Student Practice

9. Graph $y = \log_2 x$ and $y = 2^x$ on the same set of axes.

Make a table of values (ordered pairs) for each equation.

x	y
-1	$\frac{1}{2}$
0	1
1	2
2	4

x	y
$\frac{1}{2}$	-1
1	0
2	1
4	2

↑   ↑   ↑   ↑

Coordinates of ordered pairs are reversed

Graph the ordered pairs.

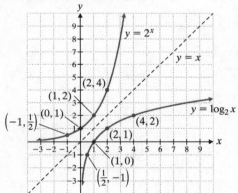

Note that $y = \log_2 x$ is the inverse of $y = 2^x$ because the ordered pairs (x, y) are reversed. The sketch of the two equations shows that they are inverses.

Recall that in function notation, f^{-1} means the inverse function of f. Thus, if we write $f(x) = \log_2 x$, then $f^{-1}(x) = 2^x$.

10. Graph $y = \log_4 x$ and $y = 4^x$ on the same set of axes.

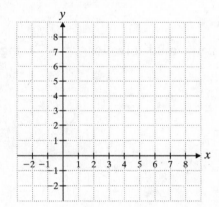

Copyright © 2013 Pearson Education, Inc.

Extra Practice

1. Write in logarithmic form. $0.0001 = 10^{-4}$

2. Write in exponential form. $-6 = \log_e x$

3. Evaluate. $\log_{10} \sqrt{10}$

4. Graph. $\log_2 x = y$

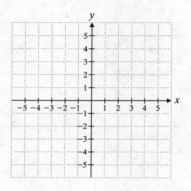

Concept Check

Explain how you would solve for x. $\quad -\dfrac{1}{2} = \log_e x$

Copyright © 2013 Pearson Education, Inc.

Chapter 11 Logarithmic and Exponential Functions
11.3 Properties of Logarithms

Vocabulary
logarithms • Logarithm of a Product Property • Logarithm of a Quotient Property
Logarithm of a Number Raised to a Power Property

1. The _____ states that for any positive real numbers M and N and any positive base $b \neq 1$, $\log_b MN = \log_b M + \log_b N$.

2. The _____ is similar to the logarithm of a product property, except that it involves two expression that are divided.

3. _____ are used to reduce complex expressions to addition and subtraction.

Example	Student Practice
1. Write $\log_3 XZ$ as a sum of logarithms.	**2.** Write $\log_5 BC$ as a sum of logarithms.
Use the logarithm of a product property, $\log_b MN = \log_b M + \log_b N$, to rewrite the logarithm.	
$b = 3$, $M = X$, and $N = Z$	
$\log_3 XZ = \log_3 X + \log_3 Z$	
3. Write $\log_3 16 + \log_3 x + \log_3 y$ as a single logarithm.	**4.** Write $\log_2 x + \log_2 20 + \log_2 z$ as a single logarithm.
If we extend our rule, we have $\log_b MNP = \log_b M + \log_b N + \log_b P$.	
Use this to rewrite the logarithm.	
$b = 3$, $M = 16$, $M = x$, and $P = y$	
$\log_3 16 + \log_3 x + \log_3 y = \log_3 16xy$	

Vocabulary Answers: 1. logarithm of a product property 2. logarithm of a quotient property 3. logarithms

Copyright © 2013 Pearson Education, Inc.

Example	Student Practice
5. Write $\log_3\left(\dfrac{29}{7}\right)$ as the difference of two logarithms.	**6.** Write $\log_{10}\left(\dfrac{19}{5}\right)$ as the difference of two logarithms.

5. (continued)

To rewrite the logarithm, use the logarithm of a quotient property,

$$\log_b\left(\frac{M}{N}\right) = \log_b M - \log_b N.$$

$$\log_3\left(\frac{29}{7}\right) = \log_3 29 - \log_3 7$$

Example	Student Practice
7. Express $\log_b 36 - \log_b 9$ as a single logarithm.	**8.** Express $\log_b 6 - \log_b 24$ as a single logarithm.

7. (continued)

Use the reverse of the property

$$\log_b\left(\frac{M}{N}\right) = \log_b M - \log_b N \text{ to write}$$

the difference as a single logarithm.

$$\log_b 36 - \log_b 9 = \log_b\left(\frac{36}{9}\right) = \log_b 4$$

Example	Student Practice
9. Write $\dfrac{1}{3}\log_b x + 2\log_b w - 3\log_b z$ as a single logarithm.	**10.** Write $5\log_b y - \dfrac{1}{4}\log_b w + 3\log_b z$ as a single logarithm.

9. (continued)

Use the property $\log_b M^p = p\log_b M$ to eliminate the coefficients of the logarithmic terms. Then, combine the sum of the logarithms and then the difference to obtain a single logarithm.

$$\frac{1}{3}\log_b x + 2\log_b w - 3\log_b z$$

$$= \log_b x^{1/3} + \log_b w^2 - \log_b z^3$$

$$= \log_b x^{1/3}w^2 - \log_b z^3$$

$$= \log_b\left(\frac{x^{1/3}w^2}{z^3}\right)$$

Copyright © 2013 Pearson Education, Inc.

Example	Student Practice
11. Write $\log_b\left(\dfrac{x^4 y^3}{z^2}\right)$ as a sum or difference of logarithms.	**12.** Write $\log_b\left(\dfrac{w^6 x^5}{y^3}\right)$ as a sum or difference of logarithms.

11. (continued)

Use the logarithm of a quotient property to rewrite the logarithm as a difference.

$$\log_b\left(\frac{x^4 y^3}{z^2}\right) = \log_b x^4 y^3 - \log_b z^2$$

Now, use the logarithm of a product property.

$$\log_b x^4 y^3 - \log_b z^2$$
$$= \log_b x^4 + \log_b y^3 - \log_b z^2$$

Finally, use the logarithm of a number raised to a power property to eliminate the exponents and get the following result.

$$4\log_b x + 3\log_b y - 2\log_b z$$

13. Evaluate.	**14.** Evaluate.
(a) Evaluate $\log_7 7$.	**(a)** Evaluate $\log_{20} 20$.
Since $\log_b b = 1$, $\log_7 7 = 1$.	
(b) Evaluate $\log_5 1$.	**(b)** Evaluate $\log_7 1$.
Since $\log_b 1 = 0$, $\log_5 1 = 0$.	
(c) Find x if $\log_3 x = \log_3 17$.	**(c)** Find x if $\log_5 x = \log_5 12$.
If $\log_b x = \log_b y$, then $x = y$.	
Thus, since $\log_3 x = \log_3 17$, $x = 17$.	

Copyright © 2013 Pearson Education, Inc.

Example	Student Practice
15. Find x if $2\log_7 3 - 4\log_7 2 = \log_7 x$.	**16.** Find x if $2\log_{11} 7 - 3\log_{11} 5 = \log_{11} x$.

Use the logarithm of a number raised to a power property to eliminate the coefficients of the logarithmic terms.

$$\log_7 3^2 - \log_7 2^4 = \log_7 x$$
$$\log_7 9 - \log_7 16 = \log_7 x$$

Then, use the logarithm of a quotient property and then the property that if $\log_b x = \log_b y$, then $x = y$, to find x.

$$\log_7 \left(\frac{9}{16}\right) = \log_7 x$$
$$\frac{9}{16} = x$$

Extra Practice

1. Write as a sum or difference of logarithms. $\log_7 x^3 y z^2$

2. Write as a single logarithm.
$\frac{1}{3}\log_a 6 + 2\log_a 6 - 5\log_a x$

3. Use the properties of logarithms to simplify the following. $\frac{1}{5}\log_3 3 - \log_{11} 1$

4. Find x if $\log_4 x + \log_4 2 = 3$.

Concept Check

Explain how you would simplify $\log_{10}(0.001)$.

Copyright © 2013 Pearson Education, Inc.

Chapter 11 Logarithmic and Exponential Functions
11.4 Common Logarithms, Natural Logarithms, and Change of Base Logarithms

Vocabulary

common logarithm • antilogarithm • natural logarithm • change of base formula

1. Base 10 logarithms are called _____ and are usually written with no subscripts.

2. The _____ is $\log_b x = \dfrac{\log_a x}{\log_a b}$, where a, b, and $x > 0$, $a \neq 1$, and $b \neq 1$.

3. Logarithms with base e are known as _____ and are usually written $\ln x$.

Example	**Student Practice**
1. On a scientific calculator or a graphing calculator, find a decimal approximation for each of the following.	**2.** On a scientific calculator or a graphing calculator, find a decimal approximation for each of the following.
(a) $\log 7.32$	**(a)** $\log 12.67$
Enter the number 7.32 on a calculator and then press the log key.	
$\log 7.32 \approx 0.864511081$	
(b) $\log 73.2$	**(b)** $\log 126.7$
$\log 73.2 \approx 1.864511081$	
3. Find an approximate value for x if $\log x = 4.326$.	**4.** Find an approximate value for x if $\log x = 7.438$.
Find the antilogarithm. We know that $\log_{10} x = 4.326$ is equivalent to $10^{4.326} = x$. Solve this problem by finding the value of $10^{4.326}$. Using a calculator, we have the following.	
$x \approx 21{,}183.61135$	

Vocabulary Answers: 1. common logarithm 2. change of base formula 3. natural logarithms

Copyright © 2013 Pearson Education, Inc.

Example	Student Practice
5. Evaluate antilog(-1.6784). This is the equivalent to asking what the value is of $10^{-1.6784}$. Be sure to enter the negative sign on your calculator. $10^{-1.6784} \approx 0.020970076$	**6.** Evaluate antilog(-2.5041).
7. On a scientific calculator, approximate the following values. **(a)** $\ln 7.21$ This is asking for the natural log of 7.21. Use the $\boxed{\ln}$ key on your calculator to solve. $\ln 7.21 \approx 1.975468951$ **(b)** $\ln 72.1$ $\ln 72.1 \approx 4.278054044$	**8.** On a scientific calculator, approximate the following values. **(a)** $\ln 3.25$ **(b)** $\ln 55.93$
9. On a scientific calculator, find an approximate value for x for each equation. **(a)** $\ln x = 2.9836$ If $\ln x = 2.9836$, then $e^{2.9836} = x$. Use the $\boxed{e^x}$ key on a scientific calculator to solve. Thus, $x = e^{2.9836} \approx 19.75882051$. **(b)** $\ln x = -1.5619$ If $\ln x = -1.5619$, then $e^{-1.5619} = x$. Thus, $x = e^{-1.5619} \approx 0.209737192$.	**10.** On a scientific calculator, find an approximate value for x for each equation. **(a)** $\ln x = 6.0123$ **(b)** $\ln x = -0.9774$

Copyright © 2013 Pearson Education, Inc.

Example	Student Practice
11. Evaluate using common logarithms. $\log_3 5.12$	**12.** Evaluate using common logarithms. $\log_6 7.981$

11. Evaluate using common logarithms. $\log_3 5.12$

Use the change of base formula. Here $b = 3$ and $x = 5.12$.

$$\log_b x = \frac{\log_a x}{\log_a b}$$

$$\log_3 5.12 = \frac{\log 5.12}{\log 3} \approx 1.486561234$$

The answer is approximate with nine decimal places, depending on the calculator you may have more or fewer digits.

13. Obtain an approximate value for $\log_4 0.005739$ using natural logarithms.

Use the change of base formula where $a = e$, $b = 4$ and $x = 0.005739$.

$$\log_4 0.005739 = \frac{\log_e 0.005739}{\log_e 4}$$

$$= \frac{\ln 0.005739}{\ln 4}$$

$$\approx -3.722492455$$

To check, we want to know the following.

$$4^{-3.722492455} \overset{?}{=} 0.005739$$

Using a calculator, this can be verified using the $\boxed{y^x}$ key. The answer checks.

14. Obtain an approximate value for $\log_2 0.02546$ using natural logarithms.

Copyright © 2013 Pearson Education, Inc.

Example	Student Practice
15. Using a scientific calculator, graph $y = \log_2 x$.	**16.** Using a scientific calculator, graph $y = \log_4 x$.

Use the change of base formula with common logarithms to find $y = \dfrac{\log x}{\log 2}$.

Find values of y for various values of x and organize them into a data table. Graph the data points.

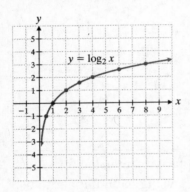

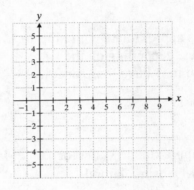

Extra Practice

1. On a scientific calculator or a graphing calculator, find a decimal approximation for $\log 11.2$.

2. On a scientific calculator, find an approximate value for x if $\ln x = 3.2$.

3. Evaluate using common logarithms. $\log_{12} 0.451$

4. Obtain an approximate value for $\log_4 0.0332$ using natural logarithms.

Concept Check

Explain how you would find x using a scientific calculator if $\ln x = 1.7821$.

Copyright © 2013 Pearson Education, Inc.

Chapter 11 Logarithmic and Exponential Functions
11.5 Exponential and Logarithmic Equations

Vocabulary

logarithm   •   natural logarithm   •   logarithmic equation   •   exponential equation

1. If an exponential equation involves e raised to a power, take the _____ of each side of the equation.

2. It is not possible to take the _____ of a negative number.

3. To solve a(n) _____, get one logarithmic term on one side and a numerical value on the other, then convert it to an exponential equation using the definition of a logarithm.

Example	Student Practice
1. Solve. $\log_3(x+6) - \log_3(x-2) = 2$	**2.** Solve. $\log_2(x+4) - \log_2(x-2) = 2$

Apply property 2, which states that $\log_b \dfrac{M}{N} = \log_b M - \log_b N$. Then write the equation in exponential form.

$$\log_3(x+6) - \log_3(x-2) = 2$$

$$\log_3\left(\frac{x+6}{x-2}\right) = 2$$

$$\frac{x+6}{x-2} = 3^2$$

Solve the resulting equation.

$$\frac{x+6}{x-2} = 9$$

$$x+6 = 9(x-2)$$

$$x+6 = 9x-18$$

$$24 = 8x$$

$$3 = x$$

The check is left to the student.

Vocabulary Answers: 1. natural logarithm 2. logarithm 3. logarithmic equation

Copyright © 2013 Pearson Education, Inc.

Example	Student Practice
3. Solve. $$\log(x+6)+\log(x+2)=\log(x+20)$$	**4.** Solve. $$\log(x+6)+\log(x-4)=\log(3x-4)$$

3. (continued)

Use property 6, if $b \neq 1$, $x > 0$, $y > 0$ and $\log_b x = \log_b y$, then $x = y$.

$$\log(x+6)+\log(x+2)=\log(x+20)$$

$$\log(x+6)(x+2)=\log(x+20)$$

$$x^2+8x+12=x+20$$

$$x^2+7x-8=0$$

$$(x+8)(x-1)=0$$

$$x-1=0 \quad \text{or} \quad x+8=0$$
$$x=1 \qquad\qquad x=-8$$

Check the solutions.

$$\log(1+6)+\log(1+2)\overset{?}{=}\log(1+20)$$

$$\log(7\cdot3)\overset{?}{=}\log 21$$

$$\log 21 = \log 21$$

$$\log(-8+6)+\log(-8+2)\overset{?}{=}\log(-8+20)$$

$$\log(-2)+\log(-6)\neq\log(12)$$

Discard -8 because it leads to taking the logarithm of a negative number. Thus, the only solution is $x=1$.

5. Solve $2^x = 7$. Leave your answer in exact form.	**6.** Solve $5^x = 12$. Leave your answer in exact form.

5. (continued)

Take the logarithm of each side. Solve for x.

$$2^x = 7$$

$$\log 2^x = \log 7$$

$$x\log 2 = \log 7$$

$$x = \frac{\log 7}{\log 2}$$

Copyright © 2013 Pearson Education, Inc.

Example	Student Practice
7. Solve $e^{2.5x} = 8.42$. Round your answer to the nearest ten-thousandth.	**8.** Solve $e^{4.7x} = 10.75$. Round your answer to the nearest ten-thousandth.

$$\ln e^{2.5x} = \ln 8.42$$
$$(2.5x)(\ln e) = \ln 8.42$$
$$2.5x = \ln 8.42$$
$$x = \frac{\ln 8.42}{2.5}$$
$$x \approx 0.8522$$

---

9. If P dollars are invested in an account that earns interest at 12% compounded annually, the amount available after t years is $A = P(1+0.12)^t$. How many years will it take for \$300 in this account to grow to \$1500? Round your answer to the nearest whole year.

Substitute the known values and simplify.

$$1500 = 300(1+0.12)^t$$
$$1500 = 300(1.12)^t$$
$$5 = (1.12)^t$$

Now, take the common logarithm of each side and solve for t.

$$\log 5 = \log(1.12)^t$$
$$\log 5 = t(\log 1.12)$$
$$\frac{\log 5}{\log 1.12} = t$$
$$14.20150519 \approx t$$

Thus, it would take approximately 14 years.

10. If P dollars are invested in an account that earns interest at 15% compounded annually, the amount available after t years is $A = P(1+0.15)^t$. How many years will it take for \$150 in this account to grow to \$1700? Round your answer to the nearest whole year.

Copyright © 2013 Pearson Education, Inc.

Example	Student Practice
11. At the beginning of 2011, the world population was seven billion people and the growth rate was 1.2% per year. If this growth rate continues, how many years will it take for the population to double to fourteen billion?	**12.** At the beginning of 2011, the mosquito population in a certain county was eight million and the growth rate was 1.7% per year. If this growth rate continues, how many years will it take for the population to double to sixteen million?

Use the formula $A = A_0 e^{rt}$ and write the population in terms of billions.

$$A = A_0 e^{rt}$$
$$14 = 7 e^{(0.012)t}$$
$$2 = e^{(0.012)t}$$
$$\ln 2 = \ln e^{(0.012)t}$$
$$\ln 2 = 0.012t$$
$$57.76226505 \approx t$$

It would take about 58 years.

Extra Practice

1. Solve. $\log 2x + \log 4 = \log(2x + 12)$

2. Solve. $\ln 8 - \ln x = \ln(x - 7)$

3. Solve $6^x = 4^{x+2}$. Round your answer to the nearest thousandth.

4. If P dollars are invested in an account that earns interest at 6% compounded annually, the amount available after t years is $A = P(1 + 0.06)^t$. How many years will it take for $4000 in this account to grow to $7000? Round your answer to the nearest whole year.

Concept Check
Explain how you would solve $26 = 52e^{3x}$.

282

Copyright © 2013 Pearson Education, Inc.

MATH COACH

Mastering the skills you need to do well on the test.

Watch the **MATH COACH** videos in MyMathLab® or on You Tube™ while you work the problems below. These helpful hints will help you avoid making common errors on test problems.

Solving an Exponential Equation—Problem 3

Solve. $4^{x+3} = 64$

> **Helpful Hint:** If one side of the equation is in exponential form, it is best to try to write the other side of the equation in exponential form. Then the procedure will be easier to complete.

Did you rewrite the equation as $4^{x+3} = 4^3$?
Yes _____ No _____

If you answered No, remember to write your equation in the form $b^x = b^y$. Note that $64 = 4^3$, and you want the base of the exponent on each side of the equation to be the same number, b, such that $b > 0$ and $b \neq 1$.

Did you use the property of exponential equations to write the equation $x + 3 = 3$?
Yes _____ No _____

If you answered No, remember that if $b^x = b^y$, then $x = y$ for any $b > 0$ and $b \neq 1$. Now solve the equation for x.

If you answered Problem 3 incorrectly, go back and rework the problem using these suggestions.

Using the Properties of Logarithms to Write Sums and Differences of Logarithms as a Single Logarithm—Problem 6
Write as a single logarithm. $2\log_7 x + \log_7 y - \log_7 4$

> **Helpful Hint:** Try your best to memorize the three properties of logarithms in Objectives 11.3.1, 11.3.2, and 11.3.3. They are essential to know when working with logarithmic expressions.

Did you use property 3 to eliminate the coefficient of the first logarithmic term, $2\log_7 x$, and rewrite that term as $\log_7 x^2$?
Yes _____ No _____

If you answered No, review property 3 in Objective 11.3.3 and complete this step again.

Did you use property 1 to combine the sum of the first two logarithmic terms and obtain the expression,

$\log_7 x^2 y - \log_7 4$?
Yes _____ No _____

If you answered No, review property 1 in Objective 11.3.1 and complete this step again.

In your last step, you will need to use property 2 in 11.3.2 to combine the difference of two logarithms.

Now go back and rework the problem using these suggestions.

Copyright © 2013 Pearson Education, Inc.

Solving a Logarithmic Equation—Problem 12

Solve the equation and check your solution. $\log_8(x+3) - \log_8 2x = \log_8 4$

> **Helpful Hint:** Use the properties of logarithms to rewrite the equation such that one logarithmic term appears on each side of the equation.

First, did you combine the two logarithms on the left side of the equation and rewrite the equation as

$$\log_8\left(\frac{x+3}{2x}\right) = \log_8 4?$$

Yes ____ No ____

If you answered No, use property 2 from Objective 11.3.2 to complete this first step again.

Next, did you rewrite the equation as $\dfrac{x+3}{2x} = 4$?

Yes ____ No ____

If you answered No, notice that you now have one logarithmic term on each side of the equation, which is the goal mentioned in the Helpful Hint. You can use property 6 from Objective 11.3.4 to evaluate these two logarithms.

In your final step, solve the equation for x. Check your solution by substituting for x in the original equation, evaluating the logarithms and then simplifying the resulting equation.

If you answered Problem 12 incorrectly, go back and rework the problem using these suggestions.

Solving an Exponential Equation Involving e Raised to a Power—Problem 14

Solve the equation. Leave your answer in exact form. Do not approximate. $e^{5x-3} = 57$

> **Helpful Hint:** The first step is to take the natural logarithm of each side of the equation. Then you can simplify the equation further using the properties of logarithms.

Did you take the natural logarithm of each side of the equation to obtain $\ln e^{5x-3} = \ln 57$?
Yes ____ No ____

Next did you rewrite this equation as $(5x-3)(\ln e) = \ln 57$?
Yes ____ No ____

If you answered No to these questions, review property 3 of logarithms from Objective 11.3.3 and complete this step again.

Did you rewrite the equation as $5x - 3 = \ln 57$?
Yes ____ No ____ .

If you answered No, remember that $\ln e = \log_e e = 1$. This combines the definition of natural logarithms or $\ln e = \log_e e$ and property 4 of logarithms from Objective 11.3.4, which indicates that $\log_e e = 1$.

In your final step, solve the equation for x without evaluating $\ln 57$.

Now go back and rework the problem using these suggestions.

Copyright © 2013 Pearson Education, Inc.

1.1
Student Practice

2. (a) Rational number and real number
    (b) Rational number and real number
    (c) Integer, rational number, and real number
    (d) Irrational number and real number
4. (a) Commutative property of addition
    (b) Identity property of addition
6. (a) Associative property of multiplication
    (b) Distributive property of multiplication over addition
    (c) Inverse property of multiplication

Extra Practice

1. $-33, \; -\dfrac{48}{5}, \; -2.347, \; -0.222...,$

    $0, \; \dfrac{1}{8}, \; \sqrt{81}, \; 43.5$
2. $0, \; \sqrt{81}$
3. Associative property of multiplication
4. Identity property of addition

Concept Check

Answers may vary. Possible solution:
$(4)(3.5+9.3)=(4)(9.3+3.5)$
illustrates the commutative property of addition. The statement
$(4)(9.3+3.5)=(9.3+3.5)(4)$
illustrates the commutative property of multiplication.

1.2
Student Practice

2. (a) $\dfrac{1}{9}$
    (b) 5.7

    (c) 13
4. (a) -4.5
    (b) $\dfrac{8}{15}$
6. (a) -13
    (b) $\dfrac{1}{4}$
8. (a) -25
    (b) $-\dfrac{17}{20}$
    (c) 4.6
10. $-\dfrac{15}{16}$
12. $\dfrac{4}{27}$
14. 42
16. -4

Extra Practice

1. 4
2. $\dfrac{1}{35}$
3. $-\dfrac{4}{3}$
4. 0

Concept Check

Answers may vary. Possible solution:
Simplify the numerator and denominator separately. Then divide.
$$\dfrac{3(-2)+8}{5-9}=\dfrac{-6+8}{-4}=\dfrac{2}{-4}=-\dfrac{1}{2}$$

1.3
Student Practice

2. a^6
4. (a) $-\dfrac{1}{64}$
    (b) -4096
6. The square roots of 64 are 8 and -8. The principal square root of 64 is 8.

Copyright © 2013 Pearson Education, Inc.

8. (a) 12
   (b) 2
   (c) −6
10. (a) 0.7
   (b) $\dfrac{9}{10}$
   (c) This is not a real number.
12. −35
14. 155
16. 20

Extra Practice
1. $-\dfrac{1}{32}$
2. −0.4
3. 210
4. 4

Concept Check
Answers may vary. Possible solution:
Evaluate the numerator first.

$$\sqrt{(-3)^3 - 6(-2) + 15} = \sqrt{-27 - 6(-2) + 15}$$
$$= \sqrt{-27 + 12 + 15}$$
$$= \sqrt{0}$$
$$= 0$$

Then evaluate the denominator.
$$|3 - 5| = |-2| = 2$$

Since the numerator is 0 (and the denominator is not), the expression evaluates to 0.

1.4
Student Practice
2. $\dfrac{1}{b^7}$
4. $-21x^5y^4$
6. (a) −6
   (b) 1
   (c) $\dfrac{-3}{z^4}$
8. $\dfrac{1}{6xy^{13}}$

10. $\dfrac{9}{4x^4 y^4 z^6}$
12. $-\dfrac{25y^{12}}{216x^5}$
14. (a) 976.3
   (b) 0.000001112
16. 4.0×10^{-5}

Extra Practice
1. $6a^7 b^6$
2. $\dfrac{1}{2a^{12}}$
3. $\dfrac{64x^{15}}{y^3}$
4. 3.2×10^{-5}

Concept Check
Answers may vary. Possible solution:
Multiply the constants. Then simplify the variables by adding exponents. Rewrite the answer using only positive exponents.

$$\left(3x^2 y^{-3}\right)\left(2x^4 y^2\right) = 3 \cdot 2x^{2+4} y^{-3+2}$$
$$= 6x^6 y^{-1}$$
$$= \dfrac{6x^6}{y}$$

1.5
Student Practice
2. (a) $8z^4,\ -3y$
   (b) $3z,\ 4w,$ and 7
4. Coefficient of the x^2 term is $\dfrac{5}{6}$, coefficient of the z term is −2.5, coefficient of the p^3 term is 18,000.
6. (a) $-2a^2 + a - 11$
   (b) $10.1z^3 - 5.6z^2 - 2.3z$
8. $-3a^4 - 12a^3$
10. (a) $-12a^2 - 4a^3 - 8a$
   (b) $-12y^2 z^2 + 18yz^4 - 6y^3 z^2 + 24yz^2$
12. (a) $-2y^3 - 3y^2 + y - 7$

Copyright © 2013 Pearson Education, Inc.

(b) $14a^2 - 21ab$

14. $a - 18b$

16 $3a - 10b$

Extra Practice

1. $3x^3 - 2x$

2. $-y^3 + 3y^2 - y$

3. $\dfrac{2}{3}x^4 - x^3 + 3x$

4. $8y^3 - 2y + 18$

Concept Check

Answers may vary. Possible solution:
Combine like terms. $-3x$ and $-8x$ are
like terms. $4y$ and $-5y$ are like terms.

$2x^2 - 3x + 4y - 2x^2 y - 8x - 5y$

$= 2x^2 + (-3 - 8)x + (4 - 5)y - 2x^2 y$

$= 2x^2 - 11x - y - 2x^2 y$

1.6

Student Practice

2. 12

4. -88

6. (a) 576

   (b) -144

8. $-49°F$

10. \$2530

12. 60 ft

14. 80 ft^2

16. 904.32 in.^{3}

Extra Practice

1. 132

2. 3

3. \$2232

4. 28.26 in.

Concept Check

Answers may vary. Possible solution:
Substitute $p = 5000$, $r = 0.08$, and
$t = 2$ into the formula and simplify.

$A = p(1 + rt)$

$= 5000\left[1 + (0.08)(2)\right]$

$= 5000(1 + 0.16)$

$= 5000(1.16) = 5800$

Copyright © 2013 Pearson Education, Inc.

2.1

Student Practice

2. Yes

4. $x = -24$

6. $x = \dfrac{10}{7}$

8. $x = -4$

10. $x = 4$

12. No solution

14. Any real number is a solution.

Extra Practice

1. $x = 4$

2. $y = -\dfrac{13}{9}$

3. $x = 45$

4. $x = -1$

Concept Check

Answers may vary. Possible solution:
Multiply both sides by the LCD,
$2 \cdot 3 \cdot 5 = 30$. Then simplify by
obtaining all numerical values on one
side and all variable terms on the
other side. Divide both sides by -21.

$$\frac{3x+1}{2} + \frac{2}{3} = \frac{4x}{5}$$

$$30\left(\frac{3x+1}{2} + \frac{2}{3}\right) = 30\left(\frac{4x}{5}\right)$$

$$15(3x+1) + 10(2) = 6(4x)$$

$$45x + 15 + 20 = 24x$$

$$45x + 35 = 24x$$

$$45x - 45x + 35 = 24x - 45x$$

$$35 = -21x$$

$$\frac{35}{-21} = \frac{-21x}{-21}$$

$$-\frac{5}{3} = x$$

2.2

Student Practice

2. $y = \dfrac{2 - 5x}{3}$

4. $z = \dfrac{5A - 4x - 4y}{8}$

6. $z = \dfrac{-5ax - 10a}{12a - 8}$

8. $x = \dfrac{970 - 10t}{3}$; 2015

10. (a) $\dfrac{A}{ar} + 1$

(b) $\dfrac{9}{4}$ or $2\dfrac{1}{4}$

Extra Practice

1. $x = \dfrac{7y}{5} + 21$

2. $t = \dfrac{I}{pr}$

3. $y = \dfrac{-x}{w - 2}$

4. (a) $h = \dfrac{S - 2lw}{2w + 2l}$

(b) 3 inches

Concept Check

Answers may vary. Possible solution:
Isolate all terms containing x on one
side of the equals sign. Then divide
both sides by the coefficient of x.

$$3(3ax + y) = 2ax - 5y$$

$$9ax + 3y = 2ax - 5y$$

$$9ax - 2ax = -5y - 3y$$

$$7ax = -8y$$

$$\frac{7ax}{7a} = \frac{-8y}{7a}$$

$$x = -\frac{8y}{7a}$$

Copyright © 2013 Pearson Education, Inc.

2.3

Student Practice

2. $x = 3,\ x = -6$

4. $x = -5,\ x = -25$

6. $x = 1,\ x = -2$

8. $x = -2,\ x = -\dfrac{2}{3}$

Extra Practice

1. $x = 11,\ x = -20$

2. $m = -2,\ m = \dfrac{2}{3}$

3. $s = -\dfrac{11}{5},\ s = -\dfrac{7}{3}$

4. $x = -2,\ x = -\dfrac{32}{11}$

Concept Check

Answers may vary. Possible solution:
Since $|x| = x$ if $x \geq 0$ and $|x| = -x$ if $x < 0,$ write two separate equations by removing the absolute value bars and letting the right side of the equation equal $\dfrac{1}{2}$ and $-\dfrac{1}{2}.$ Then simplify each.

$$|2x + 4| = \frac{1}{2}$$

$2x + 4 = \dfrac{1}{2}$ $\quad$ or $\quad$ $2x + 4 = -\dfrac{1}{2}$

$\quad 2x = -4 + \dfrac{1}{2}$ $\qquad$ $2x = -4 - \dfrac{1}{2}$

$\qquad 2x = -\dfrac{7}{2}$ $\qquad\qquad$ $2x = -\dfrac{9}{2}$

$\qquad\quad x = -\dfrac{7}{4}$ $\qquad\qquad$ $x = -\dfrac{9}{4}$

2.4

Student Practice

2. 525 minutes

4. 30 weeks

6. $l = 35$ yards, $w = 42$ yards

Extra Practice

1. 76

2. 138 pages

---

3. 9 months

4. Natasha $= 14$ hours
   Mia $= 16$ hours
   Fredrick $= 8$ hours

Concept Check

Answers may vary. Possible solution:
Let the width be represented by the variable $x.$ Then define the length in terms of $x.$ Substitute the variable expressions in the perimeter formula, $P = 2(\text{width}) + 2(\text{length}),$ and simplify to solve for $x.$ Then substitute x back in the expressions for the width and length. Let $x = $ width, then $4x + 7 = $ length.

$$212 = 2x + 2(4x + 7)$$

$$212 = 2x + 8x + 14$$

$$212 = 10x + 14$$

$$198 = 10x$$

$$19.8 = x$$

$$4x + 7 = 4(19.8) + 7 = 86.2$$

The length is 86.2 feet and the width is 19.8 feet.

2.5

Student Practice

2. 1125 townspeople

4. $2500 at 11%, $3500 at 16%

6. 5 pounds of $2 a pound
   5 pounds of $5 a pound

8. 50 miles per hour for 3 hours
   65 miles per hour for 1 hour

Extra Practice

1. 30 apples

2. $3500

3. 90 grams of 80%
   60 grams of 55%

4. 42 miles per hour in heavy traffic
   72 miles per hour in thin traffic

Copyright © 2013 Pearson Education, Inc.

Concept Check
Answers may vary. Possible solution:
Let a variable represent one quantity,
such as $x =$ price the previous year. Then
write an equation in terms of x from the
problem. Solve the equation and state the
answer to the question asked in the
problem.

$$x + 0.12x = 39,200$$
$$1.12x = 39,200$$
$$x = 35,000$$

The price was \$35,000 the previous year.

2.6
Student Practice

2.
 (a) $\dfrac{3}{7} > \dfrac{2}{5}$

 (b) $-0.423 > -0.44$

4. $|5 - 13| > |2 - 6|$

6. (a)
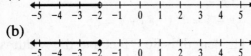

 (b)

 (c)

8.

$x \le 19$

10.

$x \ge -3$

12. $x > 2$

14. $x \le 19$

16. 15 minutes

Extra Practice

1. $-\dfrac{7}{12} < -\dfrac{5}{8}$

2.
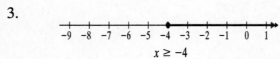

$x < -12$

3.

$x \ge -4$

4. More than 15 tables
Concept Check
Answers may vary. Possible solution:
If both sides of an inequality are multiplied
or divided by the same negative number,
the inequality symbol is reversed.
Therefore, when solving $-3x < 9$, the
inequality symbol is reversed:

$$-3x < 9$$
$$\frac{-3x}{-3} > \frac{9}{-3}$$
$$x > -3$$

2.7
Student Practice

2.

$-2 < x \text{ and } x < 2$

4.

-4.5

$4.5 < x \le 6$

6.

\$1050

$\$300 \le s \le \$1050$

8.

$x < -1 \text{ or } x > 1$

10.

$x < 21 \text{ or } x > 60$

12.

$x \le 1 \text{ or } x \ge 4$

14.

$x \ge -4 \text{ or } x < 2$

16. $\varnothing$, there is no solution

Extra Practice

1.

$\frac{1}{2}$

$-4 < x \le \frac{1}{2}$

Copyright © 2013 Pearson Education, Inc.

2.

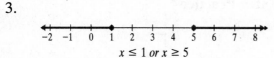

$$x \le 4 \text{ or } x \ge \frac{13}{2}$$

3.

$$x \le 1 \text{ or } x \ge 5$$

4. $\quad x < 0.6 \text{ or } x > 0.85$

Concept Check

Answers may vary. Possible solution:

$x + 8 < 3 \quad$ and $\quad 2x - 1 > 5$

$\quad x < -5 \qquad\qquad 2x > 6$

$\qquad\qquad\qquad\qquad x > 3$

Since $x < -5$ and $x > 3$ do not overlap, there are no values of x that satisfy the given conditions.

Extra Practice

1.

$$0 \le x \le 6$$

2. $\quad -5 \le x \le -1$

3.

$$x \le -3\frac{2}{3} \text{ or } x \ge 4\frac{1}{3}$$

4. $\quad 1.28 \le x \le 1.38$

Concept Check

Answers may vary. Possible solution: Since the absolute value is always nonnegative, there is no solution. $|7x + 3|$ cannot be < -4.

2.8

Student Practice

2.

$$-3 \le x \le 3$$

4.

$$-5 \le x \le 1$$

6.

$$-\frac{1}{10} < x < 1\frac{1}{10}$$

8.

$$-3 \le x \le 10$$

10.

$$x \le -4 \text{ or } x \ge 4$$

12.

$$x \le -1 \text{ or } x \ge 5$$

14.

$$x \le 1 \text{ or } x \ge 4$$

16. $\quad 273.61 \le x \le 274.35$

Copyright © 2013 Pearson Education, Inc.

Worksheet Answers Chapter 3

3.1

Student Practice

2.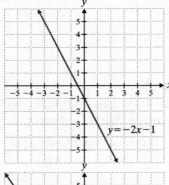

4.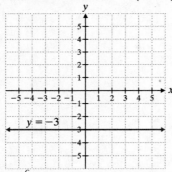

x- intercept is $(-2,0)$

y- intercept is $(0,-3)$

Additional points may vary, one possible solution is $(-4,3)$.

6.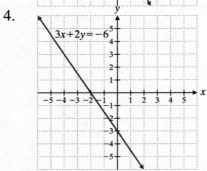

8.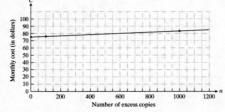

Extra Practice

1.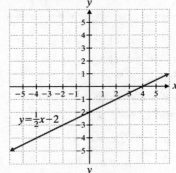

2.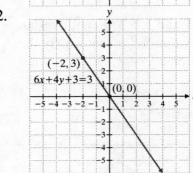

x- intercept is $(0,0)$

y- intercept is $(0,0)$

Additional points may vary, one possible solution is $(-2,3)$

3.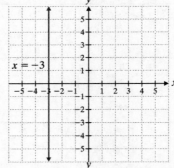

4.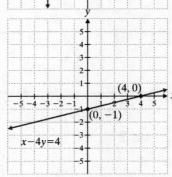

292

Copyright © 2013 Pearson Education, Inc.

x- intercept is $(4,0)$

y- intercept is $(0,-1)$

Additional points may vary, one possible solution is $(-4,-2)$.

Concept Check

Answers may vary. Possible solution:
To find the x- intercept of
$7x+3y=-14$, set $y=0$ and solve for x.

$$7x+3(0)=-14$$
$$7x=-14$$
$$x=-2$$

To find the y- intercept of
$7x+3y=-14$, set $x=0$ and solve for y.

$$7(0)+3y=-14$$
$$3y=-14$$
$$y=-4\frac{2}{3}$$

3.2

Student Practice

2. $-\frac{5}{2}$

4. (a) -2

   (b) $-\frac{25}{6}$

6. 0.6

8. $-\frac{7}{4}$

10. $m_{AB}=-2$

    $m_{BC}=-2$

    The line segments have a point (B) in common and the same slope so all three points lie on the same line.

12. Yes

Extra Practice

1. -8

2. 0.125

3. 3

4. 72 meters

Concept Check

Answers may vary. Possible solution:

$$m=\frac{\text{rise}}{\text{run}}$$

$$m=\frac{8.5}{120}$$

$$m\approx0.07 \text{ or } 7\%$$

3.3

Student Practice

2. $m=3$; y-intercept $=(0,-4)$

   $y=3x-4$

4. $m=\frac{1}{2}$; y-intercept $=(0,-2)$

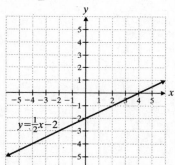
$y=\frac{1}{2}x-2$

6. $4x-y=12$

8. $y=\frac{8}{5}x-\frac{17}{5}$

10. $x-2y=3$

Extra Practice

1. $m=-1$; y-intercept $=(0,-2)$

   $y=-x-2$

2. $m=\frac{3}{5}$; y-intercept $=(0,-3)$

   $y=\frac{3}{5}x-3$

3. $y=-4x+21$

4. $5x-y=17$

Concept Check

Answers may vary. Possible solution:
Because the line is horizontal the slope is

Copyright © 2013 Pearson Education, Inc.

0, the y-intercept is $(0, -8)$. Substitute these values into the slope-intercept form.

$$y = 0(x) - 8$$
$$y = -8$$

3.4
Student Practice

2.

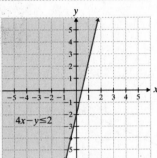

4.

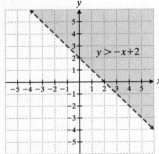

6.

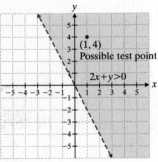

8.

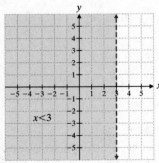

1.

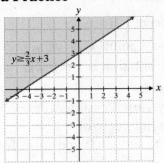

2.

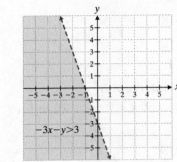

3.

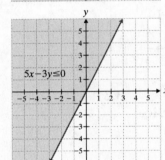

4.

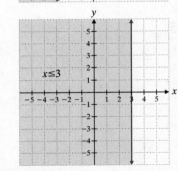

Concept Check
Answers may vary. Possible solution: First simplify the inequality.

$$4y - 8 < 0$$
$$4y < 8$$
$$y < 2$$

Graph a dashed line at $y = 2$ and shade the region below it. The solution is the

Copyright © 2013 Pearson Education, Inc.

shaded region below the dashed line, but not including the dashed line.

3.5
Student Practice

2.   Answers may vary.
     Table:

Year	2000	2001	2002	2003	2004
Time in Minutes	9.52	8.08	7.75	7.05	6.42

Ordered pairs:

$\{(2000, 9.52), (2001, 8.08),$

$(2002, 7.75), (2003, 7.05),$

$(2004, 6.42)\}$

Graph:

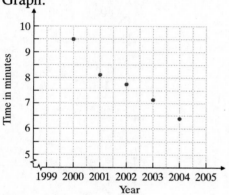

4. $\text{Domain} = \{4, 10, 8\}$

$\text{Range} = \{1, 3, 5\}$

The relation is not a function.

6. $\text{Domain} = \{5, 6, 7, 8\}$

$\text{Range} = \{36, 38, 40, 44\}$

The relation is a function.

8.   (a) The relation is a function.
     (b) The relation is not a function.

10. (a) -2
     (b) 14

Extra Practice

1. $\text{Domain} = \{5, 6, 7, 8\}$

$\text{Range} = \{23, 24, 25, 26\}$

The relation is a function.

2.   The graph does not represent a function.

3. -7

4.   8

Concept Check
Answers may vary. Possible solution:
If a vertical line can intersect the graph of a relation more than once, the relation is not a function. If no such line can be drawn, the relation is a function.

3.6
Student Practice

2.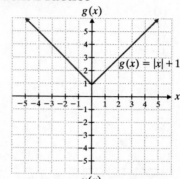
     $g(x) = |x| + 1$

4.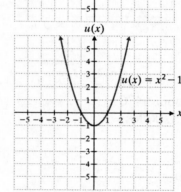
     $u(x) = x^2 - 1$

6.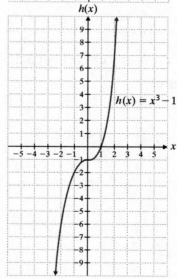
     $h(x) = x^3 - 1$

Copyright © 2013 Pearson Education, Inc.

8.

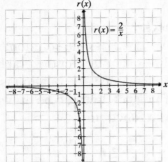

2.

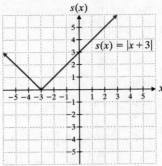

10. (a)

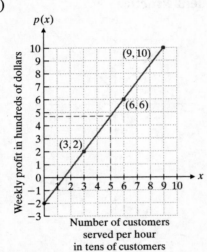

Weekly profit in hundreds of dollars

Number of customers
served per hour
in tens of customers

(b) About $470

(c) A $200 loss

3.

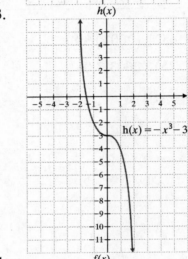

12.

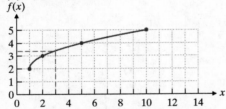

When $x = 3$, $f(x)$ is about 3.5.

Extra Practice

1.

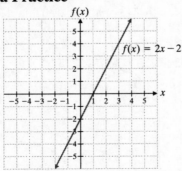

4.

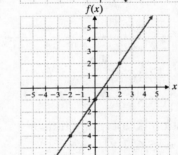

Concept Check

Answers may vary. Possible solution:

To evaluate for $f\left(\dfrac{1}{4}\right)$, substitute all

values of x with $\dfrac{1}{4}$.

$$f(x) = \frac{5}{8x - 5}$$

$$f\left(\frac{1}{4}\right) = \frac{5}{8\left(\dfrac{1}{4}\right) - 5} = -\frac{5}{3}$$

Copyright © 2013 Pearson Education, Inc.

4.1
Student Practice

2. $(3,2)$ is a solution

4. $(4,-2)$

6. $(10,3)$

8. $(4,5)$

10. No solution; inconsistent system

12. Infinite number of solutions; dependent system

14. (a) $(650,1585)$

    (b) $(4,-2)$

Extra Practice

1. $(-1,2)$

2. No solution; inconsistent system

3. $(0,4)$

4. Infinite number of solutions; dependent system

Concept Check

Answers may vary. Possible solution: When the addition method is used, the result of the addition, $0=0,$ is an identity. The system is dependent and has an infinite number of solutions.

4.2
Student Practice

2. $(2,8,-7)$ is a solution.

4. $(4,8,-2)$

6. $(-1,2,-5)$

Extra Practice

1. $(-2,-2,5)$

2. $(-1,-1,1)$

3. $(6,6,1)$

4. No solution; inconsistent system

Concept Check

Answers may vary. Possible solution: Add 5 times the first equation to 2 times the second equation to eliminate z. Then, add -3 times the second equation to 5 times the third equation to eliminate z. The resulting equations will be in terms of x and y only.

4.3
Student Practice

2. 250 advance tickets; 125 door tickets

4. Boat speed is 15 miles per hour; current speed is 5 miles per hour

6. Eight 12-ton trucks; five 8-ton trucks; three 6-ton trucks

Extra Practice

1. 71, 31

2. 185 acres of soybeans, 315 acres of corn

3. 5 binders, 7 pens, and 3 erasers

4. 122 adults, 80 students, 48 senior citizens

Concept Check

Answers may vary. Possible solution: The equations would be set up the same except the right sides would be set to 1500 rather than 1200.

4.4
Student Practice

2.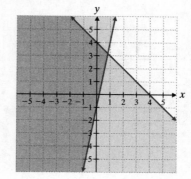

Copyright © 2013 Pearson Education, Inc.

4.

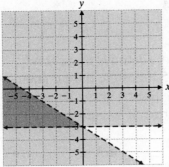

3.

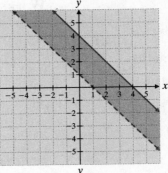

6.

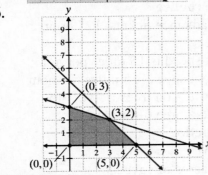

4.

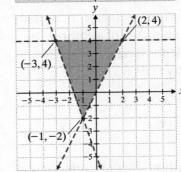

Extra Practice

1.

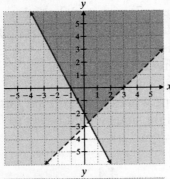

2.

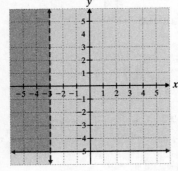

Concept Check

Answers may vary. Possible solution: $y > x + 2$ is graphed using a dashed line and shaded above the line. $x < 3$ is graphed using a dashed line and shaded to the left of the line. The region that satisfies both inequalities is the overlap in the shaded regions.

Copyright © 2013 Pearson Education, Inc.

5.1

Student Practice

2. (a) Trinomial, degree 5
   (b) Trinomial, degree 8
   (d) Monomial, degree 10

4. (a) 54
   (b) −84

6. $2y^2 - 5y + 8$

8. $-4y^2 + 17y - 4$

10. $32x^2 + 16x - 30$

12. $9x^2 - 49y^2$

14. $16x^2 + 72xy^2 + 81y^4$

16. $-15x^3 + 44x^2 - 42x + 16$

Extra Practice

1. $-6x^2 - 3$

2. $2x^3 - 6x^2 - x + 3$

3. $16x^4 - 40x^2y + 25y^2$

4. $8a^3 - 26a^2 + 19a - 3$

Concept Check

Answers may vary. Possible solution:
The first step is to use the FOIL
method to multiply the first two
polynomials. Then combine the like
terms of the product. Next use the
distributive property to multiply the
remaining polynomial by the above
product.

5.2

Student Practice

2. $6y^3 + 4y^2 - 9$

4. $3x^2 + 4x + 1 + \dfrac{7}{4x - 1}$

6. $49x^2 + 14x + 4$

8. $3x^2 - x + 2$

Extra Practice

1. $-5x^2y^2 + 2xy + 1$

2. $x - 8 + \dfrac{24}{x + 2}$

3. $2x^2 + 3x + 2 + \dfrac{9}{2x - 3}$

4. $x^2 + 6$

Concept Check

Answers may vary. Possible solution:
Multiply the divisor by the quotient,
then add the remainder. The result will
be the dividend if the division was
performed correctly.

5.3

Student Practice

2. $5x^2 - 12x + 30 + \dfrac{-80}{x + 3}$

4. $4x^3 - 5x^2 + x + 6$

6. $5x^3 + 3x^2 + 10x + 17 + \dfrac{13}{x - 2}$

Extra Practice

1. $3x^2 - 2x + 5$

2. $3x^3 + 6x^2 + 12x + 25 + \dfrac{52}{x - 2}$

3. $x^2 + 4x + 14 + \dfrac{62}{x - 4}$

4. $2x^4 + 6x^3 - 2x^2 + 1 - \dfrac{6}{x + 2}$

Concept Check

Answers may vary. Possible solution:
When a term x^n of the dividend is
missing in the sequence of descending
powers of x, a zero must be used for
the missing term's coefficient. This is
done to save a place in the quotient for
a coefficient of the term x^{n-1}.

5.4

Student Practice

2. $9y(2 - y)$

4. $5x(3x^2 - 4xy^3 + 8y^4 - 6y)$

6. $4a^2(6a + 9b - 5b^2)$

Copyright © 2013 Pearson Education, Inc.

8. $(y-4)(6-7y)$

10. $(m+5n)(x+5z)$

12. $(x-4)(6x+5z)$

14. $(n-3)(m+7)$

16. $(x^2-7)(4x-9)$

Extra Practice

1. $3xy(10x^2-xy+5)$

2. $(x-3)(3x-4y+2)$

3. $(x-2)(3-2y)$

4. $(x-2)(x^2-3)$

Concept Check

Answers may vary. Possible solution:
Arrange the order as follows:
$8xy-10y+12x-15$.

5.5

Student Practice

2. $(x-8)(x-4)$

4. $(x-12)(x+3)$

6. $(x^2-5)(x^2+2)$

8. $(x+6y)(x-4y)$

10. $(3x+5)(x+4)$

12. $4x(2x-5)(x-2)$

14. $(3x^2+2)(2x^2-5)$

Extra Practice

1. $(x-7)(x-2)$

2. $(x+8)(x-1)$

3. $2(x+3y)(x-2y)$

4. $(7x+5)(2x-3)$

Concept Check

Answers may vary. Possible solution:
The first step in factoring the expression
$3x^2y^2+6xy^2-72y^2$ is to factor $3y^2$ out
of every term.
$3x^2y^2+6xy^2-72y^2=3y^2(x^2+2x-24)$

5.6

Student Practice

2. $(x+5)(x-5)$

4. $(13x^2+4y^2)(13x^2-4y^2)$

6. $(3x-4)^2$

8. $2(11x+7)^2$

10. $(6x^2+7y^2)^2$

12. $(4x+3y)(16x^2-12xy+9y^2)$

14. $(6x^2-7y^3)(36x^4+42x^2y^3+49y^6)$

16. $3(4x-1)(16x^2+4x+1)$

Extra Practice

1. $2x(3x+4y)(3x-4y)$

2. $(2x-3)^2$

3. $12(2x-3)^2$

4. $3(3x-2y)(9x^2+6xy+4y^2)$

Concept Check

Answers may vary. Possible solution:
The middle coefficient is only half of
what would be required for the formula
to work.

5.7

Student Practice

2. (a) $2(2x+5)(2x-5)$

   (b) $2x^2(2x-3)(4x^2+6x+9)$

4. (a) $4(5x+3y)^2$

   (b) $3(x-12)(x-3)$

   (c) $x^2(4x-3)(3x+2)$

   (d) $6(w-z)(b+7)$

6. Prime

8. Prime

Extra Practice

1. Prime

2. $2x(x-6)^2$

3. $(x^2+1)(x+1)(x-1)$

Copyright © 2013 Pearson Education, Inc.

4. $(x+3)(x+2)(x-2)$

Concept Check

Answers may vary. Possible solution:
With all signs positive, the second term
coefficient must equal 2 times the
product of the first and third
coefficient.

5.8

Student Practice

2. $x = -15,\ x = 2$

4. $x = \dfrac{4}{5}$

6. $x = 0,\ x = -4,\ x = 7$

8. Base = 5 meters
    Altitude = 12 meters

Extra Practice

1. $x = 8,\ x = 2$

2. $x = 0,\ x = -\dfrac{2}{7}$

3. Length = 10 cm, Width = 7 cm

4. 9 cm by 9 cm

Concept Check

Answers may vary. Possible solution:
Multiply the equation by the LCD (8)
to eliminate fractions. Next move all
terms to the left side of the equation
and factor to get $(3x-4)(x+2)=0$.
Set each term equal to zero and solve
for x to get 2 possible solutions of
$x = \dfrac{4}{3}$ and $x = -2$. Test the validity of
each possible solution in the context of
the problem, discard possible solutions
that make no sense.

Copyright © 2013 Pearson Education, Inc.

Worksheet Answers Chapter 6

6.1

Student Practice

2. All real numbers except 3 and -15

4. $\dfrac{12m-n}{m+n}$

6. $-\dfrac{9y-x}{8x+3y}$

8. $\dfrac{1}{2x(x+7)}$

10. $\dfrac{5(a+b)}{9ax^3}$

12. $\dfrac{2(6a+b)}{7x(a+10b)}$

Extra Practice

1. $\dfrac{x+2}{6}$

2. $-\dfrac{3y}{2y-3}$

3. $\dfrac{x^2}{x+4}$

4. $\dfrac{1}{2x+1}$

Concept Check

Answers may vary. Possible solution: The first step is to factor -1 out of the numerator.

$$\dfrac{-(x^2-9)}{x^2-7x+12}$$

The next step is to factor the numerator and the denominator.

$$\dfrac{-(x+3)(x-3)}{(x-4)(x-3)}$$

Lastly, remove common factors.

$$-\dfrac{x+3}{x-4}$$

6.2

Student Practice

2. $(7x+3y)(7x-3y)$

4. $40x^4y^2$

6. $\dfrac{5x-1}{(3x+5)(x-8)}$

8. $\dfrac{14x-9}{(x+9)(x-1)}$

10. $\dfrac{8y^2+15x^2}{40x^4y^3}$

12. $\dfrac{-6x^2+10-3}{3(3x+7)^2}$

Extra Practice

1. $x(3x-1)(x+2)$

2. $\dfrac{-4x+1}{x^2-11x+6}$

3. $\dfrac{3(3y-2)}{(y-1)(y+1)(y-2)}$

4. $\dfrac{x+12}{(x-3)(5x-2)}$

Concept Check

Answers may vary. Possible solution: Factor both denominators into prime factors. List all the different prime factors. The LCD is the product of these factors, each of which is raised to the highest power that appears on that factor in the denominators.

6.3

Student Practice

2. $\dfrac{m^2(m^2+8)}{2(m+2)}$

4. $\dfrac{(5y-24)(y+5)}{7(y^2-5y+1)}$

6. $\dfrac{a}{10-a}$

Copyright © 2013 Pearson Education, Inc.

8. $\dfrac{5x(3y+5)}{(2y+5)(9y-7x)}$

Extra Practice

1. $\dfrac{x^6}{4y^4}$

2. $\dfrac{3(x+5)}{x-3}$

3. $-\dfrac{2(6+a+b)}{3a}$

4. $-\dfrac{y+10}{5}$

Concept Check
Answers may vary. Possible solution:
First find the LCD of all the fractions in the numerator and denominator. The LCD is xy. Next multiply the numerator and denominator by the LCD. Use the distributive property. The result will be simplified, but should be written in factored form.

6.4

Student Practice

2. $x=3$
4. $x=5$
6. $z=-4$
8. No solution
10. No solution

Extra Practice

1. $x=-2$
2. $x=2$
3. $x=\dfrac{1}{4}$
4. No solution

Concept Check
Answers may vary. Possible solution:
The left side of the equation equals zero. Therefore, it cannot equal $\dfrac{3}{2}$. It has no solution.

6.5

Student Practice

2. $b=\dfrac{af}{a-f}$

4. $G=\dfrac{Fd^2}{m_1 m_2}$

6. 150 managers, 850 workers
8. The helicopter is about 354 feet above the building.

Extra Practice

1. $r=\dfrac{A-P}{Pt}$

2. $g=\dfrac{s-s_0-v_0 t}{t^2}$

3. 55 inches, 70 inches
4. 2.4 hours or 2 hours and 24 minutes

Concept Check
Answers may vary. Possible solution:
To eliminate fractions, multiply both sides of the equation by the denominator $B-H$. To isolate H, first subtract SB from both sides of the equation, then divide both sides by $-S$.

Copyright © 2013 Pearson Education, Inc.

Worksheet Answers Chapter 7

7.1

Student Practice

2. $\dfrac{27}{64}a^3b^{-18}$

4. (a) $x^{12/5}$
    (b) $x^{3/4}$
    (c) $6^{9/13}$

6. (a) $-12x^{9/10}$
    (b) $5x^{7/8}y^{-2/3}$

8. $-8x^{7/6}+24x^{1/2}$

10. (a) 64
    (b) 16

12. $\dfrac{5+x}{x^{3/4}}$

14. $6z\left(3z^{2/3}-4z^{1/3}\right)$

Extra Practice

1. $x^{16/5}$

2. $ab^{4/3}$

3. $\dfrac{1+3y}{y^{2/3}}$

4. $5x\left(2x^{3/4}+5x^{1/8}\right)$

Concept Check

Answers may vary. Possible solution:
Change the exponents to have equal
denominators, then add the
numerators over the common
denominator. This is the combined,
simplified exponent for x.

7.2

Student Practice

2. (a) 4
    (b) 1
    (c) 3.3

4. The domain is all real numbers x,
    where $x \geq 9$.

6. 7

8. (a) $z^{3/5}$
    (b) $a^{8/3}$

10. (a) $\sqrt[3]{a^5b^5}$ or $\sqrt[3]{(ab)^5}$
    (b) $\sqrt[5]{5x}$

12. (a) $\dfrac{1}{4}$
    (b) Not a real number

14. (a) $5|a|$
    (b) $4z^2$
    (c) $3x^2|y|$

Extra Practice

1. 7

2. $(4m-3n)^{5/8}$

3. $\dfrac{1}{\left(\sqrt[7]{4}\right)^4}$ or $\dfrac{1}{\sqrt[7]{256}}$

4. $7|a^3|b^8$

Concept Check

Answers may vary. Possible solution:
Factor the coefficient completely to
identify the fourth root. Remove the
fourth root from under the radical.
Divide the exponents of the variables
by 4 to remove the variables from
under the radical.

7.3

Student Practice

2. $5\sqrt{2}$

4. $6\sqrt{3}$

6. $-3\sqrt[3]{6}$

8. (a) $4a^2b^2\sqrt{2a}$
    (b) $4yz\sqrt[3]{3x^2yz^2}$

10. $4\sqrt{3a}$

12. $4\sqrt{5}$

14. $5\sqrt{a}-3\sqrt{2a}$

16. $14x^2z\sqrt[3]{5z}$

304
Copyright © 2013 Pearson Education, Inc.

Extra Practice

1. $5x\sqrt{x}$

2. $5b^4\sqrt[4]{ab^3}$

3. $10\sqrt{7}-4\sqrt{5}$

4. $17x\sqrt{3}$

Concept Check

Answers may vary. Possible solution: Completely factor the coefficient of the radicand to identify the fourth root. Divide the exponents of the variables by 4 to remove the variables from under the radical. The results are moved outside the radical. The remainders stay under the radical.

7.4

Student Practice

2. $-12\sqrt{21z}$

4. $39-10\sqrt{15}$

6. $12+18\sqrt{2}+8\sqrt{3}+12\sqrt{6}$

8. $11-2\sqrt{22a}+2a$

10. (a) $\dfrac{2}{3}$

    (b) $6ab^2\sqrt{a}$

12. $\dfrac{5\sqrt{2z}}{8z}$

14. $\dfrac{3\sqrt[3]{a^2}}{a}$

16. $\dfrac{31+7\sqrt{35}}{13}$

Extra Practice

1. $15a\sqrt{ab}$

2. $265+30\sqrt{70}$

3. $\dfrac{3\sqrt{2x}}{5y^2}$

4. $x\sqrt{5}+x\sqrt{3}$

Concept Check

Answers may vary. Possible solution: To rationalize the denominator, the numerator and the denominator must be multiplied by the conjugate of the denominator. In this case, the conjugate of the denominator is $3\sqrt{2}+2\sqrt{3}$. The product will not contain a radical in the denominator.

7.5

Student Practice

2. $x=6$

4. $x=-\dfrac{1}{3}$ or $x=3$

6. $x=5$

8. $y=3$

Extra Practice

1. $x=3$
2. No solution
3. $x=0$ or $x=4$
4. $x=5$

Concept Check

Answers may vary. Possible solution: Substitute the found values for x back into the original equation to test for validity.

7.6

Student Practice

2. (a) $i\sqrt{41}$

    (b) $5i$

4. -18

6. $x=-3,\ y=4\sqrt{2}$

8. $-3+2i$

10. $8-27i$

12. $-24-12i$

14. (a) -1

    (b) 1

16. $\dfrac{9-38i}{25}$ or $\dfrac{9}{25}-\dfrac{38}{25}i$

Extra Practice

1. $20-3i\sqrt{2}$
2. $-1+i$
3. $4+7i$
4. $\dfrac{30+12i}{29}$

Copyright © 2013 Pearson Education, Inc.

Concept Check
Answers may vary. Possible solution:
Multiply by the FOIL method.
Replace i^2 with -1. Combine like terms.

7.7

Student Practice

2. 144 feet

4. $y = \dfrac{25}{6}$

6. 19 lumens

8. $y = \dfrac{175}{6}$

Extra Practice

1. 40 inches

2. $y \approx 3.9$

3. 80 mph

4. $y \approx 4.2$

Concept Check
Answers may vary. Possible solution:
First write the equation as $y = \sqrt{x}k$.
Substitute the known values for x and y and solve for k.

Copyright © 2013 Pearson Education, Inc.

Worksheet Answers Chapter 8

8.1

Student Practice

2. 5 or −5
4. $2\sqrt{6}$ or $-2\sqrt{6}$
6. 4 or −4
8. $5i$ or $-5i$
10. $\dfrac{-8+\sqrt{6}}{3}$ or $\dfrac{-8-\sqrt{6}}{3}$
12. $-2+\sqrt{3}$ or $-2-\sqrt{3}$
14. $\dfrac{2+\sqrt{19}}{5}$ or $\dfrac{2-\sqrt{19}}{5}$

Extra Practice

1. $8i$ or $-8i$
2. $\dfrac{-1+\sqrt{15}}{3}$ or $\dfrac{-1-\sqrt{15}}{3}$
3. −5 or −7
4. $\dfrac{1+i\sqrt{47}}{8}$ or $\dfrac{1-i\sqrt{47}}{8}$

Concept Check

Answers may vary. Possible solution: Divide the coefficient of x (1) by 2 to get $\dfrac{1}{2}$. Since $\left(\dfrac{1}{2}\right)^2 = \dfrac{1}{4}$, add $\dfrac{1}{4}$ to both sides of the equation.

8.2

Student Practice

2. $-6\pm\sqrt{29}$
4. $\pm\sqrt{15}$
6. $x \approx 20$ or $x \approx 10$
8. $-1\pm\sqrt{11}$
10. $\dfrac{2\pm i\sqrt{14}}{6}$
12. Complex roots
14. $x^2 - 7x + 12 = 0$
16. $x^2 + 45 = 0$

Extra Practice

1. $\dfrac{-6\pm\sqrt{15}}{3}$
2. $\dfrac{-7\pm\sqrt{89}}{2}$
3. One rational solution
4. $2x^2 + 7x + 3$

Concept Check

Answers may vary. Possible solution: Subtract 3 from both sides to put the equation in standard form. Identify the values of a, b, and c, and find the value of the discriminant, $b^2 - 4ac$. If the discriminant is a perfect square, there are two rational solutions; if it is a positive number that is not a perfect square, there are two irrational solutions; if it is zero, there is one rational solution; if it is negative, there are two nonreal complex solutions.

8.3

Student Practice

2. $x = \pm4$, $x = \pm5i$
4. $x = 3$, $x = \dfrac{\sqrt[3]{18}}{3}$
6. $x = 64$, $x = 1$
8. $x = 625$
10. $x = \dfrac{1}{3}$, $x = -\dfrac{1}{4}$

Extra Practice

1. $x = \pm i\sqrt{7}$, $x = \pm i\sqrt{3}$
2. $x = 0$, $x = -3$
3. $x = -1$, $x = -32$
4. $x = \dfrac{1}{7}$, $x = -\dfrac{1}{8}$

Concept Check

Answers may vary. Possible solution: Substitute y for x^4, y^2 for x^8 which yields $y^2 - 6y = 0$. Next factor out the

Copyright © 2013 Pearson Education, Inc.

y from the left side of the equation yielding $y(y-6)=0.$ Set each term equal to zero and solve for y yielding $y=0$ and $y=6.$ Substitute x^4 for y and solve for x yielding $x=0$ and $x=\pm\sqrt[4]{6}.$

8.4

Student Practice

2. $r=\sqrt[3]{\dfrac{3V}{4\pi}}$

4. $x=2y$ or $x=-5y$

6. $x=\dfrac{3y\pm3\sqrt{y^2-4z}}{4}$

8. (a) $c=\sqrt{a^2+b^2}$
   (b) 25

10. 10 mi, 24 mi, and 26 mi

12. width $=7$ yd; length $=12$ yd

Extra Practice

1. $t=\sqrt{\dfrac{2s}{g}}$

2. $x=\dfrac{7\pm\sqrt{79-12ay-24y}}{2a+4}$

3. $a=\sqrt{10};\ b=3\sqrt{10}$

4. 25 mph, then 45 mph

Concept Check

Answers may vary. Possible solution: Set one leg's length $=x,$ then the other leg's length $=3x.$

Use $c^2=a^2+b^2$ with $c=12,\ a=x,$ and $b=3x.$ Solve for x to find one leg length then multiply the found value of x by 3 to find the other leg's length.

8.5

Student Practice

2. vertex $=(3.5,-0.25);$
   y- intercept $=(0,12);$
   x- intercepts $=(4,0),(3,0)$

4. vertex $=(2,-3);$
   y- intercept $=(0,1);$
   x- intercepts $\approx(3.7,0),(0.3,0)$

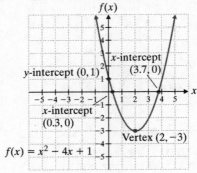

6. vertex $=(-1,-3);$
   y- intercept $=(0,-6);$
   No x- intercepts.

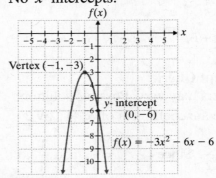

Extra Practice

1. vertex $=(-2.5,-12.25);$
   y- intercept $=(0,-6);$
   x- intercepts $=(1,0),(-6,0)$

2. vertex $=(-1.7,-26.45);$
   y- intercept $=(0,-12);$
   x- intercepts $=(0.6,0),(-4,0)$

Copyright © 2013 Pearson Education, Inc.

3.

$$\text{vertex} = \left(\frac{1}{3}, \frac{2}{3}\right) \text{ or } \approx (0.3, 0.7)$$

$y\text{-intercept} = (0,1);$

No x-intercepts

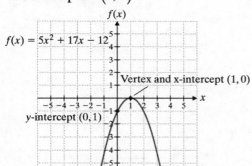

4.

$$\text{vertex} = (1,0);$$

$y\text{-intercept} = (0,-1);$

$x\text{-intercepts} = (1,0)$

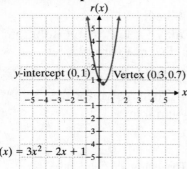

Concept Check

Answers may vary. Possible solution:
The function is in standard form,

$f(x) = ax^2 + bx + c$ with $a = 4$, $b = -9$,

and $c = -5$. The x-coordinate of the vertex

is $x_{\text{vertex}} = \dfrac{-b}{2a} = \dfrac{-(-9)}{2(4)} = \dfrac{9}{8}$. The

y-coordinate of the vertex is

$$f(x_{\text{vertex}}) = f\left(\frac{9}{8}\right).$$

Student Practice

2.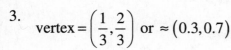

$$x < -3 \text{ or } x > 4$$

4.

$$-3 \le x \le 2.5$$

6.

$$x < -4.2 \text{ or } x > 0.2$$

Extra Practice

1.

$$-1 \le x \le 2.5$$

2. $x < -4$ or $x > 16$

3. $x = $ all real numbers

4. $x = $ all real numbers

Concept Check

Answers may vary. Possible solution:
The equation $x^2 + 2x + 8 = 0$ does not
have any real solutions. Thus, there are
no boundary points. Also, the quadratic
function $f(x) = x^2 + 2x + 8$ has no

x-intercepts, so the graph lies entirely
above or below the x-axis. This means
that the inequality is either true for all
real numbers, or has no real number
solutions.

Copyright © 2013 Pearson Education, Inc.

Worksheet Answers Chapter 9

9.1

Student Practice

2. $\sqrt{29}$

4. Center at $(1,4)$; radius is $r = 4$

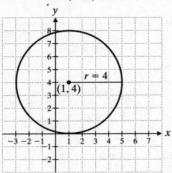

6. $(x-14)^2 + (y+5)^2 = 7$

8. $(x+2)^2 + (y-1)^2 = 9$; Center at $(-2,1)$; radius is $r = 3$

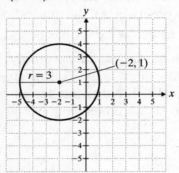

Extra Practice

1. $2\sqrt{5}$

2. $x^2 + \left(y - \dfrac{6}{5}\right)^2 = 13$

3. Center at $(0,0)$; radius at $r = 6$

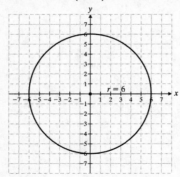

4. $(x+5)^2 + (y-3)^2 = 5$; center at $(-5,\ 3)$; radius at $r = \sqrt{5}$

Concept Check

Answers may vary. Possible solution: Using the distance formula, let $(x_1, y_1) = (-6,8)$, $(x_2, y_2) = (x,12)$, and $d = 4$. Then, solve for the unknown variable x.

9.2

Student Practice

2. Vertex at $(-1,0)$; axis of symmetry is $x = -1$

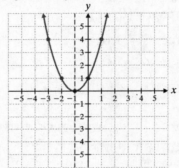

4.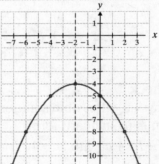

6.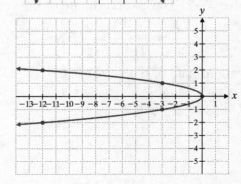

310

Copyright © 2013 Pearson Education, Inc.

8. $x = (y+3)^2 - 2$

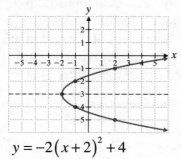

10. $y = -2(x+2)^2 + 4$

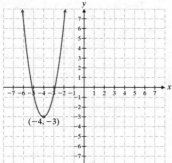

Extra Practice

1. Vertex at $(-4, -3)$; y-intercept at $(0, 29)$

2. Vertex at $\left(\dfrac{3}{2}, 3\right)$; y-intercept at $(0, -1)$

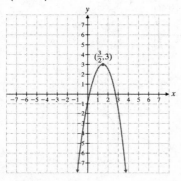

3. Vertex at $(-3, -1)$; x-intercept at $(0, 0)$

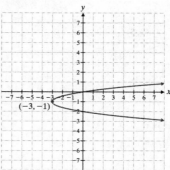

4. $x = -3(y-2)^2 + 18$

(a) Horizontal

(b) Opens left

(c) $(18, 2)$

Concept Check

Answers may vary. Possible solution:

$y = ax^2$, vertical, opens up

$y = -ax^2$, vertical, opens down

$x = ay^2$, horizontal, opens right

$x = -ay^2$, horizontal, opens left

Using these rules you can tell which way the parabolas will open.

9.3

Student Practice

2.

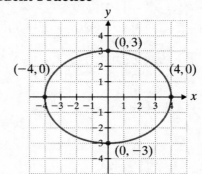

Copyright © 2013 Pearson Education, Inc.

4.

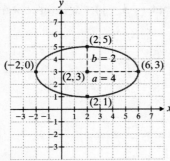

Extra Practice

1.

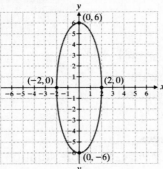

2.

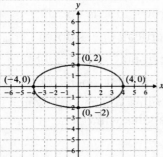

3.

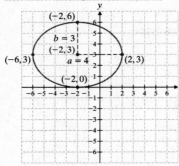

4.

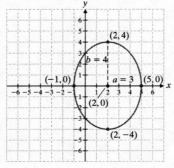

Concept Check

Answers may vary. Possible solution: Add 36 to both sides of the equation, then divide both sides of the equation by 36 to put the equation in standard form. The center of the ellipse is at $(0,0)$ so y-intercepts are $\left(0,\pm\dfrac{b}{2}\right)$ and x-intercepts are $\left(\pm\dfrac{a}{2},0\right)$.

9.4

Student Practice

2.

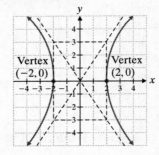

4.

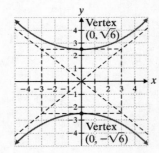

6.

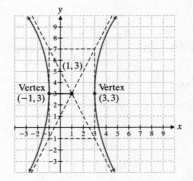

312

Copyright © 2013 Pearson Education, Inc.

Extra Practice

1.

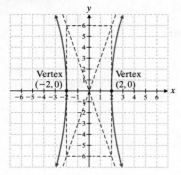

2.

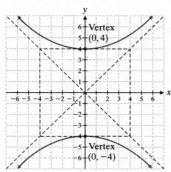

3.

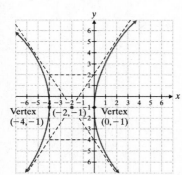

4.

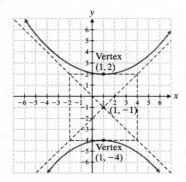

Concept Check

Answers may vary. Possible solution: Divide both sides of the equation by 196 in order to put the equation in standard form. This equation describes a horizontal hyperbola, the asymptote of

which is $y = \dfrac{b}{a}x$. Substitution yields

$y = \dfrac{7}{2}x$.

9.5
Student Practice

2. $(0,5)$, $(-7,12)$

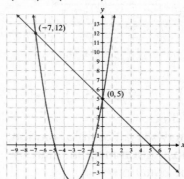

4. $\left(\dfrac{\sqrt{2}}{2}, -\sqrt{2}\right)$, $\left(-\dfrac{\sqrt{2}}{2}, \sqrt{2}\right)$

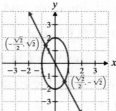

6. $\left(\dfrac{\sqrt{10}}{10}, \dfrac{\sqrt{10}}{10}\right)$, $\left(\dfrac{\sqrt{10}}{10}, -\dfrac{\sqrt{10}}{10}\right)$,

$\left(-\dfrac{\sqrt{10}}{10}, \dfrac{\sqrt{10}}{10}\right)$, $\left(-\dfrac{\sqrt{10}}{10}, -\dfrac{\sqrt{10}}{0}\right)$

Extra Practice

1. $(3,5)$, $(-2,0)$

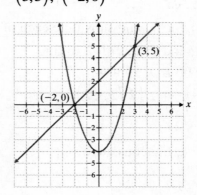

313

Copyright © 2013 Pearson Education, Inc.

2. $\left(\dfrac{29}{4}, -\dfrac{3}{4}\right)$

3. $(3,4),\ (3,-4),\ (-3,4),\ (-3,-4)$

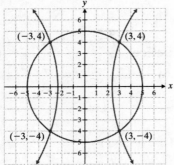

4. $(3,2),\ (3,-2),\ (-3,2),\ (-3,-2)$

Concept Check

Answers may vary. Possible solution: Use the substitution method. Start by labeling the equations.

$$y^2 + 2x^2 = 18 \quad (1)$$
$$xy = 4 \quad (2)$$

Because (2) is a linear equation, and because the y^2 term of (1) has a coefficient of 1, choose to solve (2) for the variable y and substitute the found value for y into (1).

$$y = \dfrac{4}{x} \quad (2)$$

$$\left(\dfrac{4}{x}\right)^2 + 2x^2 = 18 \quad (2) \text{ and } (1)$$

Solve the resulting equation, in terms of x only, for x.

$$x = \pm 1,\ \pm 2\sqrt{2}$$

Solve (2) for y, and substitute each of the four found values of x to find corresponding y values.

$$(1,4),\ (-1,-4),\ \left(2\sqrt{2}, \sqrt{2}\right),$$
$$\left(-2\sqrt{2}, -\sqrt{2}\right)$$

Check for extraneous answers.

314
Copyright © 2013 Pearson Education, Inc.

10.1

Student Practice

2.   (a) $6b - 7$

    (b) $6b + 29$

    (c) $6b + 22$

4. $$\dfrac{-10}{(a+8)(a+6)}$$

6.   5

8.   (a) 452.16

    (b) $S(e)$

       $= 452.16 + 150.72e + 12.56e^2$

    (c) 498.51 cm^2, the surface area
      is to large by approximately
      46.35 cm^2.

Extra Practice

1. $5a^2 + 2a + 6$

2. $2\sqrt{a^2 + 2}$

3. $\dfrac{5}{2}$

4. $8x + 4h$

Concept Check

Answers may vary. Possible solution:

For the function $k(x) = \sqrt{3x+1}$

evaluated at $k(2a-1)$, substitute

$2a - 1$ for x in the function and solve.

$k(2a-1) = \sqrt{3(2a-1)+1} = \sqrt{6a-2}$

10.2

Student Practice

2.   (a) A function

    (b) Not a function

4.

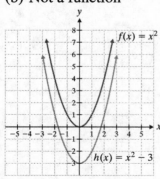

6.

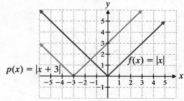

8.

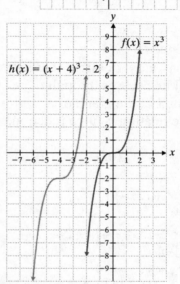

Extra Practice

1.   Not a function

2.

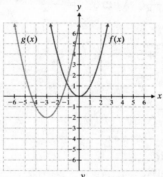

3.

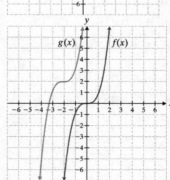

Copyright © 2013 Pearson Education, Inc.

4.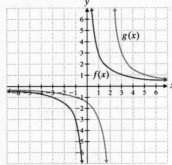

Concept Check
Answers may vary. Possible solution:
If a vertical line passes through more
than one point of the graph of a
relation, the relation is not a function.

10.3
Student Practice

2. (a) $2x^2 + 12x - 6$
   (b) 74

4. (a) $4x^3 - 15x^2 + 41x - 24$
   (b) -198

6. (a) $\dfrac{2x+3}{x-4}$, where $x \neq 4$

   (b) $\dfrac{1}{2x+3}$, where $x \neq -\dfrac{3}{2}$

8. $15x - 17$

10. (a) $\sqrt{4x+5}$
    (b) $4\sqrt{x+6} - 1$

12. (a) $\dfrac{4}{2x-5}$
    (b) -1

Extra Practice

1. (a) $1.6x^3 + 4.6x^2 - 2.7x + 7.6$
   (b) $1.6x^3 - 4.6x^2 - 2.7x - 7.6$
   (c) 84.1
   (d) -33.4

2. (a) $x^3 - 12x^2 + 48x - 64$
   (b) $x - 4$
   (c) -1
   (d) -7

3. $1 - 2x$

4. $\left| -4x - \dfrac{5}{3} \right|$

Concept Check
Answers may vary. Possible solution:
Evaluate both functions for -4, then
subtract the results of $g(-4)$ from the
results of $f(-4)$.

10.4
Student Practice

2. (a) Not one-to-one
   (b) One-to-one

4. (a) Not one-to-one
   (b) One-to-one

6. $Q^{-1} = \{(4,2),(2,6),(7,9),(3,11)\}$.

8. $f^{-1}(x) = \dfrac{x-8}{5}$

10. $f^{-1}(x) = \dfrac{x-75}{3}$

12.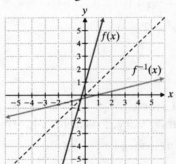

Extra Practice

1. One-to-One

2. $f^{-1}(x) = x + 4$

3. $f^{-1}(x) = \dfrac{5}{3x} + \dfrac{4}{3}$ or $\dfrac{5+4x}{3x}$

4. $g^{-1} = -\dfrac{x+4}{3}$

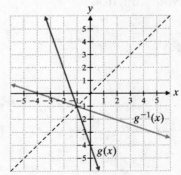

Copyright © 2013 Pearson Education, Inc.

Concept Check

Answers may vary. Possible solution:

To find the inverse of the function

$f(x) = \dfrac{x-5}{3}$, substitute y for $f(x)$.

$y = \dfrac{x-5}{3}$

Interchange x and y.

$x = \dfrac{y-5}{3}$

Solve for y in terms of x.

$x = \dfrac{y-5}{3}$

$3x = y - 5$

$y = 3x + 5$

Replace y with $f^{-1}(x)$.

$f^{-1}(x) = 3x + 5$

Copyright © 2013 Pearson Education, Inc.

11.1

Student Practice

2.

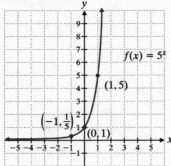

$f(x) = 5^x$

$(1, 5)$

$\left(-1, \frac{1}{5}\right)$

$(0, 1)$

4.

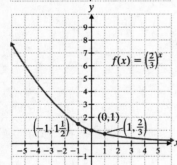

$f(x) = \left(\frac{2}{3}\right)^x$

$(0, 1)$

$\left(-1, 1\frac{1}{2}\right)$

$\left(1, \frac{2}{3}\right)$

6.

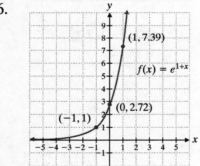

$(1, 7.39)$

$f(x) = e^{1+x}$

$(0, 2.72)$

$(-1, 1)$

8. $x = 3$

10.   \$31,990.00

12.   1.34 mg

Extra Practice

1.

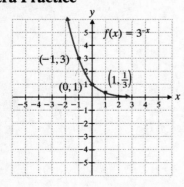

$f(x) = 3^{-x}$

$(-1, 3)$

$(0, 1)$

$\left(1, \frac{1}{3}\right)$

2.

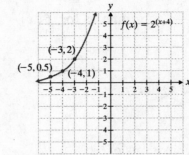

$f(x) = 2^{(x+4)}$

$(-3, 2)$

$(-5, 0.5)$ $(-4, 1)$

3. $x = -1$

4.   \$5008.62

Concept Check

Answers may vary. Possible solution:

To solve $4^{-x} = \dfrac{1}{64}$ for x, remember

that $64 = 4^{-3}$. Replace the right side of
the equation with 4^{-3}, $4^{-x} = 4^{-3}$.
This yields the same base on both
sides of the equation, and allows the
use of the property of exponential
functions to simplify, which states that
the exponents may be isolated and
compared. Thus, $-x = -3$ or $x = 3$.

11.2

Student Practice

2. $-3 = \log_6 \dfrac{1}{216}$

4. $243 = 3^5$

6.   (a) $x = \dfrac{1}{256}$

   (b) $b = 4$

8.   6

10.

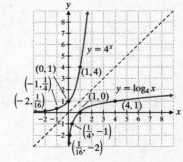

$y = 4^x$

$(0, 1)$ $(1, 4)$

$\left(-1, \frac{1}{4}\right)$ $(1, 0)$ $y = \log_4 x$

$\left(-2, \frac{1}{16}\right)$ $(4, 1)$

$\left(\frac{1}{4}, -1\right)$

$\left(\frac{1}{16}, -2\right)$

Copyright © 2013 Pearson Education, Inc.

Extra Practice

1. $\log_{10} 0.0001 = -4$

2. $e^{-6} = x$

3. $\dfrac{1}{2}$

4.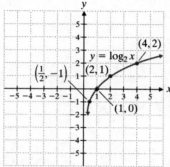

Concept Check

Answers may vary. Possible solution:

To solve $-\dfrac{1}{2} = \log_e x$ for x, write an equivalent exponential equation, $x = e^{-1/2}$.

11.3

Student Practice

2. $\log_5 B + \log_5 C$

4. $\log_2 20xz$

6. $\log_{10} 19 - \log_{10} 5$

8. $\log_b \left(\dfrac{1}{4}\right)$

10. $\log_b \left(\dfrac{y^5 z^3}{w^{1/4}}\right)$

12. $6\log_b w + 5\log_b x - \log_b y$

14. (a) 1
    (b) 0
    (c) $x = 12$

16. $x = \dfrac{49}{125}$

Extra Practice

1. $3\log_7 x + \log_7 y + 2\log_7 z$

2. $\log_a \left(\dfrac{36(6)^{1/3}}{x^5}\right)$

3. $\dfrac{1}{5}$

4. $x = 32$

Concept Check

Answers may vary. Possible solution:

To simplify $\log_{10}(0.001)$, start by setting the expression equal to x, $\log_{10}(0.001) = x$. Then, write 0.001 as a power of 10, $0.001 = 10^{-3}$. Now, rewrite the logarithm, $\log_{10} 10^{-3} = x$. Finally, use the logarithm of a number raised to a power property to write $-3\log_{10} 10 = x$, and the property $\log_b b = 1$ to get $-3(1) = x$, or $-3 = x$.

11.4

Student Practice

2. (a) 1.102776615
    (b) 2.102776615

4. $x \approx 27{,}415{,}741.72$

6. 0.003132564

8. (a) 1.178654996
    (b) 4.024100909

10. (a) 408.4216105
    (b) 0.376288177

12. 1.159231332

14. -5.295623771

16.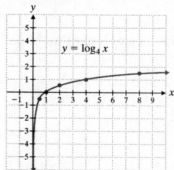

Extra Practice

1. 1.049218023

2. 24.53253019

3. -0.320449841

4. -2.456336474

Copyright © 2013 Pearson Education, Inc.

Concept Check

Answers may vary. Possible solution:
To solve $\ln x = 1.7821$ for x using a
scientific calculator, first recall that
$\ln x = 1.7821$ is equivalent to $e^{1.7821} = x$.
Then, use the calculator to evaluate
$e^{1.7821}$, resulting in $x \approx 5.942322202$.

11.5

Student Practice

2. $x = 4$
4. $x = 5$
6.
$$x = \frac{\log 12}{\log 5}$$
8. $x \approx 0.5053$
10. Approximately 17 years
12. Approximately 41 years

Extra Practice

1. $x = 2$
2. $x = 8$
3. $x \approx 6.838$
4.   Approximately 10 years

Concept Check

Answers may vary. Possible solution:
To solve $26 = 52e^{3x}$, start by dividing
both sides by 52.
$$\frac{1}{2} = e^{3x}$$
Then take the natural logarithm of both
sides.
$$\ln\left(\frac{1}{2}\right) = 3x$$
Isolate the x.
$$\frac{\ln\left(\frac{1}{2}\right)}{3} = x$$
Approximate the solution with a
calculator.
$x \approx -.0231$

Copyright © 2013 Pearson Education, Inc.